The Economics
of Environmental Policy

The Economics of Environmental Policy

A. Myrick Freeman III
Bowdoin College

Robert H. Haveman
University of Wisconsin

Allen V. Kneese
Resources for the Future, Inc.

A WILEY/HAMILTON PUBLICATION

JOHN WILEY & SONS, INC.

SANTA BARBARA • NEW YORK • LONDON • SYDNEY • TORONTO

Cover: Los Angeles County Air Pollution Control District

Library of Congress Cataloging in Publication Data:

Freeman, A. Myrick, 1936-
The economics of environmental policy.

Bibliography: p.
1. Environmental policy—United States. I. Haveman,
Robert H., joint author. II. Kneese, Allen V., joint
author. III. Title.

HC110.E5F69   301.31'0973   72-7249
ISBN 0-471-27787-8
ISBN 0-471-27786-x (pbk.)

Printed in the United States of America

10 9 8 7

Preface

This book is about the environment, economics, and institutions. Our objective is to depict the problem of environmental quality as an economic problem whose resolution requires major changes in economic, political, and legal institutions. As a consequence, we utilize knowledge and analyses developed by economists, natural scientists, political scientists, and lawyers.

In our discussion of the environment, we emphasize the principle of materials balance. This model conveys a framework for thinking systematically about the interconnectedness of the environment and the economy, and the flows of material between them. An understanding of this interconnectedness is essential to perceiving the environmental problem as an economic problem. The principle must also be recognized in developing public policy measures to induce effective and efficient reductions of the waste loads placed on the environment.

Viewed as an economic problem, the problem of environmental degradation is the result of the failure of the market system to efficiently allocate environmental resources among their alternative uses. An economic system such as ours depends primarily on voluntary exchanges of goods and services in markets to determine the relative value of things and how much of each of them to produce. This system is an effective vehicle for determining these relative values and output levels and also for efficiently allocating certain kinds of resources to the production of alternative goods and services. These resources are the ones that can readily be parceled out to individual owners who can then sell them or keep them as they wish. In an economy characterized by such privately held resources, the government's primary matter of interest, as far as resource use and allocation is concerned, is to preserve competition and assure that the distribution of income meets the society's ethical standards.

However, this market system fails to work for "common property" resources that cannot effectively be owned and traded by private individuals. The air-mantle, watercourses, and large ecological systems are examples of such resources. Since at least medieval times, it has been known that the "commons" tend to get misused and abused. Among other things, the enclosure movement of medieval England was a conservation measure. In contemporary times we are confronting the problem of the commons on an entirely new scale and we have not yet adapted our economic and political institutions to it. Much of this book delves into the nature of common property resources and their role in causing market system failure.

Building on our discussion of the environment and economics, we analyze the problem of designing institutions for collective action to efficiently manage common property environmental resources. We emphasize the need to bring environmental resources back into the economic system so that their use can be subject to the same sorts of constraints that now influence the use of other resources—land, labor, and capital. With the users of environmental resources subject to the appropriate economic incentives, progress toward improving the use of environmental resources—toward improving environmental quality—can be made.

The book presupposes no prior background in either economics or the natural sciences. Understanding it requires only a willingness to confront and master a few basic concepts—such as marginal cost and biochemical oxygen demand—and an interest in seeing how a few fairly basic economic and physical principles can help one to think more clearly about a complex set of public policy problems.

The book is appropriate for courses on economic principles or applications of economic analysis to public policy issues, as well as in survey or introductory courses in environmental problems. For the instructor with insufficient time to cover the entire book, we suggest emphasis on Chapters 2, 4, and 5. They contain the heart of our message. Chapter 2 provides the framework for viewing the environment and the economy as a single integrated system and outlines the essence of the environmental management problem. Chapter 4 discusses the role of markets in solving resource management or allocation problems. Chapter 5 outlines a model for analyzing the effect of alternative environmental management strategies. The instructor can select from among the remaining chapters according to his preferences. Chapter 3 introduces the student to some basic technical information about the nature, effects, and control of pollution. Chapters 6 and 7 deal with current public policies toward water and air pollution. Chapters 8 and 9 discuss some broader issues, such as the consistency of improved environmental quality with continued economic growth and the interrelationship of population policy with environmental policy.

We gratefully acknowledge permission to make use of the following material.

Resources for the Future, Inc., for materials appearing in Allen V. Kneese, Robert U. Ayres, and Ralph C. d'Arge, *Economics and the Environment, A Materials Balance Approach,* 1970.

The *Swedish Economic Journal,* for materials appearing in "Background for the Economic Analysis of Environmental Pollution," by Allen V. Kneese, March 1971.

The Atlantic Council of the United States and the Battelle Memorial Institute, for materials appearing in "The Economics of Environmental Management in the United States," by Allen V. Kneese in *Managing the Environment,* Allen V. Kneese, Sidney E. Rolfe, and Joseph W. Harned, editors, 1971.

General Learning Corporation, for materials appearing in "The Economics of Pollution Control and Environmental Quality," by A. Myrick Freeman III, 1971.

American Association for the Advancement of Science, for materials appearing in "Residuals Charges for Pollution Control: A Policy Evaluation," by A. Myrick Freeman III and Robert H. Haveman, in *Science,* (copyright 1972 by the American Association for the Advancement of Science).

Basic Books, Inc., for materials appearing in "Clean Rhetoric and Dirty Water," by A. Myrick Freeman III and Robert H. Haveman, in *The Public Interest,* summer 1972.

A. Alan Schmid and Marc J. Roberts read earlier drafts of the manuscript and made very helpful comments. We are grateful for their suggestions and absolve them from any responsibility for the final product. Finally, we thank Carol Gatewood for typing the final draft of the manuscript.

A. Myrick Freeman III
Robert H. Haveman
Allen V. Kneese

Contents

The Economics
of Environmental Policy

1

Environmental Quality: An Economic Problem

This book is about the environment and its quality. In writing the book, we are expressing our agreement that deterioration of the environment has become a significant national problem, but one which economics can help us to understand better and to deal with more effectively. This is partly because the basic cause of the problem is economic in nature—the failure of the market system to allocate environmental resources efficiently. Also, the choices we must make in undertaking to improve environmental quality are economic choices. They entail the attainment of economic welfare by a society—welfare that depends on how the society's resources are allocated among the many options that are available. In this first chapter, we discuss why the degradation of the environment has become a prominent national issue and how a knowledge of economics is helpful in understanding the problem and evaluating proposed solutions to it.

THE ENVIRONMENT AS A NATIONAL ISSUE

The United States economy is an incredibly productive machine. Each year it absorbs billions of tons of natural resources and turns out goods and services which we either consume or reinvest for future production. The nation's social accountants faithfully record these goods and services and value them at their market prices. The total value of this output we call the gross national product (GNP). In 1972, the economy produced over $1.1 trillion of these goods and services—over $5000 worth for each man, woman, and child in the country.

While this GNP is impressive because of both its size and the diversity of what it measures, it fails to record all of the outputs of the economic machine. As the economy is producing these goods and services that contribute to what we call our "standard of living," it is simultaneously "producing" other things—polluted rivers and streams, the smog that characterizes all of our major cities, congestion, and the encroachment on our wilderness areas —all of which detract from our "quality of life." These outputs are not recorded by the nation's social accountants even though they are, in a real sense, produced jointly along with automobiles, air conditioners, and tin cans.

One of the reasons why these effects are not recorded and counted in the GNP is that they are not bought and sold, as are the other more desirable outputs. Because they are not bought and sold, they have no price attached to them and, without a price, they are very difficult to enter into the GNP. A second reason why such undesirable outputs are not entered into the GNP concerns their importance relative to the valuable outputs. When social accounting procedures were developed in the United States in the 1920s and 1930s, the importance of these undesired outputs in the quality of life was not perceived nearly so clearly as it is now. One can only speculate about how these economic "bads" would be handled if the nation's social accounts were just now being established for the first time.

To be sure, the deleterious environmental effects of producing and consuming are more clearly perceived today than they were thirty years ago. However, we are not, as some of the frantic discussions of the environmental crisis imply, suddenly confronted with a totally new problem. Dirty rivers, dirty air, and natural resource depletion have been with us a long time, even though the pollution situation is now turning a different and, in some ways, more ominous face in our direction.

Indeed, horror stories about environmental conditions in medieval times can easily be collected. The situation in England is especially well-documented. For example, in the 14th century, London butchers had been assigned a spot at Seacoal Lane near Fleet Prison. The effect of their activity was described in a royal document as follows: "By the killing of great beasts, from whose putrid blood running down the street and the bowels cast into the Thames, the air in the city is very much corrupted and infected, whence abominable and most filthy stinks proceed, sickness and many other evils have happened to such as have abode in the said city, or have resorted to it."[1] Five centuries later—in 1850— Charles Dickens is reported to have said that "He knew of many places in it [London] unsurpassed in the accumulated horrors of

---

[1] Quoted in B. Lambert, *History and Survey of London*, Vol. 1 (London, 1806), p. 241.

their long neglect by the dirtiest old spots in the dirtiest old towns, under the worst old governments of Europe."[2]

Clearly, such abominable environmental conditions had been eliminated in most developed countries by the middle of the 20th century. Even the pervasive soot problem—caused by the use of coal furnaces to heat homes and factories and remembered well by our parents—is a thing of the past. Nevertheless, environmental degradation is now regarded by many as the nation's number one problem. What has happened to account for this? Several factors, it appears, have combined to heighten people's concern with the environment.

First, massive increases in industrial production, energy conversion, and the associated flow of materials and energy have altered the physical, chemical, and biological quality of the atmosphere and water in a truly pervasive scale. Our sensitivity to these changes has been heightened with the development of the means to detect and measure the deterioration in natural systems. Furthermore, the observation of a technology, sophisticated enough to place a man on the moon, has led to demands that it be used to improve human health or reduce the discomforts of daily life.

A second reason for the increasing environmental sensitivity relates to the "exotic" materials that are being inserted into the environment. The near alchemy of modern chemistry and physics has recently subjected the world's biological systems to strange substances to which they cannot adapt (at least, not quickly), or which are not absorbed and transformed by nature's recycling processes. The dumping of mercury and other heavy metals into the nation's rivers is one example, as is the alarmingly high observed DDT content of mothers' milk.

The third factor stems from the changing tastes of people. Ordinary people have come to expect standards of cleanliness, safety, healthfulness, and convenience that previously were the exclusive province of the wellborn or rich or simply were not imagined as possible. Moreover, with the increasing availability of leisure time, the environment has become heavily demanded for the recreational facilities and amenities that it provides.

Finally, we have become acutely aware of the tremendous increase in world population during the last few centuries. Doubts about the sustainability and consequences of continued population growth in a finite world have crept into almost everyone's mind. Indeed, neo-Malthusianism has arisen. Many people have come to feel that further applications of technology to force increased

[2] From *The Public Health as a Public Question: First Report of the Metropolitan Sanitary Association*, Address of Charles Dickens, Esq., London, 1850.

production of food and material goods from our environment is merely a holding operation against the ultimate disaster of overpopulation. Moreover, the congestion and crowding that have accompanied the increasing size and concentration of the population have themselves been viewed as an environmental problem.

SOME ECONOMIC ASPECTS OF THE ENVIRONMENTAL PROBLEM

While this discussion has served as an introduction to causes of the experienced but unmeasured problem of environmental pollution, it does not tell us why economics is relevant to understanding the nature of and relief from the problem. Perhaps the best way to establish the tie between economics and the environment is to outline several current environmental policy issues and pose some of the questions that economists ask about them—and try to answer.

Issue 1—Automotive Emission Standards

In December 1970, Congress passed legislation dealing with, among other things, air pollutant emissions from automobiles. The new law requires that automobiles sold in the 1975 model year and thereafter meet standards for their emissions of carbon monoxide and hydrocarbons that are 90 percent lower than the federal standards in effect for 1970 model cars. And, since the 1970 automotive standard already requires about a 70 percent reduction in emissions relative to prestandard automobiles, the law in effect requires a control level of up to 97 percent. During the hearings and debate on the bill, the automobile industry steadfastly maintained that it was technologically impossible to meet these standards in the time required, even taking into account the possibility of a one year extension of the 1975 deadline that the law permits.

What are some of the economic issues here? First, how does one decide that 90 percent control is better than 50 percent or 95 percent or even 100 percent? Second, what will be the costs of obtaining any of these reductions, assuming that they are in fact possible? And who will bear these costs? Official estimates place the cost of achieving these standards at between $230 and $350 per car.[3] However, these direct costs are not the only costs. There

---

[3] U.S. Environmental Protection Agency, *The Economics of Clean Air: A Report to the Congress*, Washington, D.C., February 1972.

will be many indirect costs as the economic system transmits the impact of these standards to the automotive repair industry, to the petroleum industry, and to the mass transportation industry, among other places.

What will we have bought with this expenditure? How much cleaner will our air be as a consequence of these standards? Will this improvement in the air quality be worth the cost? Finally, if it is worthwhile to achieve this air quality improvement, is there any better—less costly—way to achieve it? This question is particularly pertinent in light of a recent report from the National Air Pollution Control Administration (NAPCA) which reported that a little over one half of the 1968 model cars it tested did not pass the exhaust emissions standards established by pre-1970 federal law. Similar tests by the State of California showed that on an average, carbon monoxide emissions were about 25 percent higher than the applicable federal standards.[4] If the less stringent standards for 1968 cars are so widely violated now, how costly will it be to enforce effectively the new 1975 standards?

Issue 2—The Sulfur Oxides Tax

President Nixon has proposed a tax on emissions of sulfur oxides into the atmosphere. In his 1971 message to the Congress on the environment, he said:

"Last year in my state of the union message I urged that the price of goods 'should be made to include the cost of producing and disposing of them without damage to the environment.' A charge on sulfur emitted into the atmosphere would be a major step in applying the principle that the costs of pollution should be included in the price of the product."[5]

Is this a sound principle? And if so, why? How high should the tax be? What factors determine the appropriate level for such a tax? Who will really pay the tax—producers or the consumers? And what difference does it make? Will producers respond to the tax by introducing new production technologies, by using different fuels, or will they simply cut back on production? Finally, if the purpose of the tax is to reduce air pollution, is this the most effective policy strategy to use to obtain the goal of cleaner air?

---

[4] See U.S. Senate Committee on Public Works, Sub-committee on Air and Water Pollution—*1970 Hearings,* Washington, D.C., 1970, pp. 362–371.
[5] The President of the United States, *Message to the Congress,* Washington, D.C., 1971.

Issue 3—Federal Grants for Sewage Treatment Plant Construction

Under present legislation, the federal government makes grants to share in the costs of constructing municipal sewage treatment works. Under certain circumstances, the federal government will bear up to 55 percent of the cost of these facilities. Through fiscal year 1970 these grants had totaled about $1.4 billion. It has been officially estimated that the costs of building the additional municipal treatment facilities needed to meet our water quality standards by 1974 will total $12.2 billion.[6] The federal share of this could run to over $6 billion. In addition there is strong pressure on the federal government to make retroactive grants to reimburse municipalities that have built treatment plants without federal aid. This could entail almost a billion dollars more.

What are the objectives of the federal grant program? Clearly, one of the objectives is to accelerate the construction of municipal waste treatment facilities in order to reduce the flow of wastes to the nation's water courses. To what extent has the program attained this objective? How do we account for the fact that despite the expenditure of $5.4 billion by federal, state, and local governments over the last twelve years for waste water treatment, the quality of water in almost all of our rivers is lower today than it was twelve years ago?[7] Is this kind of federal aid the most effective means of inducing the treatment of municipal wastes?

Because of this program, the cost of constructing municipal waste treatment plants is borne by both federal taxpayers and those paying local property or sales taxes. These latter taxes are believed to be regressive—meaning that lower income people pay a higher proportion of their income in these taxes than do higher income people. What is the allocation of the burden among various income groups? Are there more equitable ways of financing this program?

Issue 4—Zero Population Growth

The zero population growth (ZPG) movement has achieved considerable momentum in this country during the decade of the 1960s. The objective of the movement is to achieve a balance

[6] U.S. Environmental Protection Agency, Water Quality Office, *Cost of Clean Water, Vol. 1*, Washington D.C., March 1971.

[7] Comptroller General of the United States, *Examination into the Effectiveness of the Construction Grant Program for Abating, Controlling and Preventing Water Pollution*, Washington, D.C., November 3, 1969.

between the birth rate and death rate in the United States. In part because of the efforts of the movement, legislation has been introduced to reduce or eliminate personal income tax exemptions for children after the second child. This would have the effect of increasing the income taxes paid by larger families, thus raising the costs of having more than two children.

Simultaneously, negative income tax (or welfare reform) legislation has been introduced and debated. Because the program provides income support to families depending on their earnings and family size, it shifts a part of the costs of additional children to the federal government. Hence, this effect works in the opposite direction as the elimination of the personal exemption proposal and, in part, offsets it.

Is either of these proposals, if they pass, likely to have much effect on population growth? Are birth rates likely to be affected by changes in the cost of bearing and rearing children? If these policies are effective in altering birth rates, what impact will they have on the future deterioration of the environment? What is the relationship between population growth and pollution levels? If changes in the economic signals (prices and costs) given to individuals can affect their decisions about family size, are there other ways in which economic incentives can be used in the fight to halt pollution?

The issues outlined here are real and important ones. The questions we have posed about them indicate that economics has something to say about all of the issues. Imagine that the ways in which economics impinges on each of these issues were jotted down on a separate piece of paper and the pieces sorted out. Upon sorting they would tend to fall rather naturally into three groups. These groups would represent the three basic economic dimensions to the environmental quality problem.

The first of these dimensions concerns the relationship between pollution and environmental degradation, on the one hand, and the economic activities of production and consumption, on the other. We are all aware by now that a connection exists between our use of cars, furnaces, and air conditioners and air pollution. It is relatively easy to make the leap from this observation to the unwarranted conclusion that the only way to halt pollution is to ban automobiles and air conditioners. However, while pollution, production, and consumption are intimately tied together, they are not produced in fixed proportions. By altering production procedures and introducing new technologies, it is possible to get more output with the same amount or less pollution. One of the tasks of economic analysis is to help us to sort out and better understand the ways in which population growth, economic growth, and growth in pollution are linked and how the current linkage can be changed. By way of preview it can be revealed that neither present population levels nor present population growth rates can be assessed the major share of the blame for our present environmental problems in the United States. We will also see that it is not economic growth per se that has caused the problem, but rather the form that this economic growth has taken. And finally we will identify ways in which we can alter the relationships between population growth, economic growth, and pollution. This knowledge will help us to formulate policies to achieve the goal of a cleaner environment.

The second economic dimension of the pollution problem relates to economic theories about human behavior. Much economic theory analyzes the behavior of economic decision makers—individuals and firms—when they are motivated by gain or profit and when they are faced with the economic signals generated by the market economy. The goal of economic theory is the explanation and prediction of behavior in these situations. This theory is required to explain why pollution occurs in a market economy. It indicates that pollution arises largely because the economic incentives or signals facing households and firms are inappropriate in that they encourage the overuse, misuse, and abuse of the environment.

This sort of economic analysis will also be useful in evaluating the public policy alternatives that are available. This is true because the objective of public policies is to modify or to control the behavior of economic units. Indeed, almost all public policies aimed at regulating human behavior operate by changing the perceived structure of incentives and rewards faced by individuals. Since pollution stems from economic activity, and arises because of distorted economic incentives, the way that the structure of incentives is affected by pollution control policies must be carefully examined. Knowledge of economics is valuable in assessing the effectiveness of policy alternatives by identifying the true structure of incentives and rewards that they embody and in determining whether they will induce behavioral changes in the desired direction and of the desired magnitude.

The third economic dimension of the pollution problem concerns choices. Economics is said to be the study of the effectiveness of the choices that must be made when unlimited wants confront scarce or limited resources. Unlimited wants means that if you could have all you wanted of everything at no cost or effort, you can always think of at least one thing you want a little more of. Scarcity means you cannot have it. When desire meets scarcity, some of the wants (probably all) do not get *completely* satisfied.

Scarce resources must be allocated or rationed among the competing wants. This allocation or rationing may be efficient in that it enables the attainment of desired objectives, or it may be inefficient. Economics is relevant for analyzing the effectiveness of such choices or allocations.

Societies face problems of choice in the face of scarcity that are similar to those confronted by an individual or a family. Such a problem of choice is confronted when we decide what level of pollution control or environmental quality we want to achieve. Viewing the problem in this way makes it clear that calls to end pollution are unrealistic. The costs of achieving complete control of pollution are probably far greater than we as a society would desire to incur. On the other hand, it seems clear that we are increasingly unwilling to live (and die) with present levels of pollution. Pollution control, like everything else, is scarce, and we must choose how much of it we want. The range of choice lies between zero pollution and zero pollution control. The choice will not be easy to make politically since it involves conflicting interests. While economic analysis cannot make these choices for us, it can illuminate the choice problem through the concepts of benefits, opportunity costs, and the optimum level of pollution control. These concepts will be discussed in later chapters.

In addition, choices must be made concerning how we are going to obtain the chosen level of environmental quality. For example, we must choose among different technological options available for controlling pollution. A technological option is any step that could be taken to reduce or control the discharge of wastes into the environment. In any given situation, some technological options will cost more per unit of pollution control than others. Thus it is important to choose these technological options that minimize the cost of obtaining a given level of pollution control. Otherwise, pollution control will cost more than necessary. We would not be getting our money's worth on each dollar spent. Moreover, with limited budgets we would get less environmental quality than we could have with the resources available.

A BRIEF PREVIEW

Having established the importance of economic understanding in solving the environmental problem, we will discuss a number of the most important economic issues relating to environmental quality. In Chapter 2, the environment will be viewed in its dual role as provider of resource inputs to the production process (as well as other valuable services) and as the receptacle of the residuals and wastes generated by production and consumption. The analysis in this chapter will hinge on the materials balance model. Chapter 3 presents some of the basic facts about environmental pollution and introduces some of the physical and technical concepts required for understanding the characteristics of the problem. This chapter also describes the extent of the problem and evaluates the direction and magnitude of the changes in environmental quality that have taken place over time.

Chapter 4 discusses the role of the market system as an allocator of resources. It emphasizes the role of prices as signals that influence resource allocation. In this context, the environmental problem is viewed as a failure of the market system because the services of the environment are not priced and exchanged as are other valuable resources.

Chapter 5 presents the more rigorous analytics of the problem. Cost and benefit functions are described and used to analyze the consequences of social decisions with respect to the environment. The notion of the optimal level of pollution control is developed and the factors that determine this optimum are analyzed.

In Chapters 6 and 7, the economic analysis developed in earlier chapters is applied to the public policy issues posed by air and water pollution. The costs and benefits of attaining improved air and water quality are discussed and the effectiveness of current and proposed policy strategies are evaluated. It is concluded that many policy approaches are deficient in that they fail to recognize the crucial role of economic incentives in attaining improved environmental quality.

Finally, Chapters 8 and 9 discuss some of the broader economic, legal, and political issues that surround the environmental management problem. The magnitude of the aggregate costs of improving the environment is assessed as is the potential role of new technologies and population control. In Chapter 9, the role of legal and political institutions in implementing sound environmental improvements is discussed. We emphasize the need to adopt a comprehensive approach to environmental management if the environmental goals of the society are to be met.

2

The Environment
and the Economy

In this chapter, we look at the environment as a provider of materials and services to the economy. In particular, we focus on the flow of materials from the environment to the economy and the return flow of these materials back to the environment as wastes or residuals. We shall see that this return flow has an adverse impact on the volume and quality of other environmental services. Viewed as a source of inputs and a receptacle for wastes, the environment is described as a resource to be managed so as to provide the maximum possible benefits to mankind. Finally, we present an economic definition of pollution and examine the relationships among pollution, economic activity, and the flow of materials to and from the environment.

THE MATERIALS BALANCE MODEL

The relationships between pollution and economic activity can be described by a simple *materials balance model* such as that portrayed in Figure 2-1. In that diagram, all of the production activities of the economic system are represented by the box labeled "production sector." Located in this box are all of the mines, factories, warehouses, transportation networks, and public utilities, for instance, that are engaged in the extraction of materials from the environment, their processing, refinement, and rearrangement into marketable goods and services, and their transportation and distribution throughout the economy to the point of ultimate use.

Whatever is produced in the production sector of this model economy goes to individuals acting as consumers. They are represented by the box labeled "household sector." Together, these two boxes make up what is usually called the economic system.

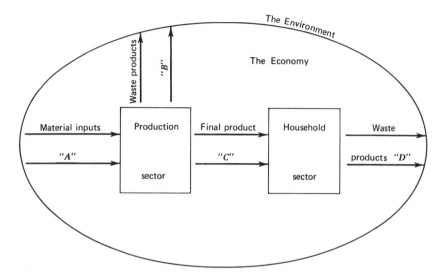

Figure 2-1 Materials balance and the economy. The materials balance for: (1) The production sector: $A = B + C$ (2) The household sector: $C = D$ (3) The economy: $A = B + D$ (flows are measured by mass).

Conventional portrayals of the economic system show a circular flow of money accompanied by an opposite flow of goods and services and productive factors between the household sector and the production sector. The household sector provides factor inputs; that is, capital and labor in return for money payments or income. In turn, the production sector provides its goods and services output to the household sector in return for money payments. Thus, the circular flow is completed.[1] However, from the perspective of the materials balance model such a view of the economic system is misleading because it ignores important flows of materials and the basic laws of physics governing them. Given that goods and services are made out of "something," the conventional model fails to indicate where that something comes from and where it goes.

[1] Some schematic representations of the economic system show the government sector as a third box. For our purposes, some of the activities of the government (for instance, national defense and building roads) can be considered as production and can be included in the production sector. Others might appropriately be labeled consumption and placed in the household sector. An example might be education. However, the distinction is unimportant for our purposes.

In the materials balance model, the environment can be viewed as a large shell surrounding the economic system. It has the same relationship to the economy as does a mother to an unborn child—it provides sustenance and carries away wastes. These input and waste flows are also portrayed in Figure 2-1. Raw materials flow from the environment, are processed in the production sector (that is, converted into consumer goods), and then—at least in part—pass on to the household sector. The materials returning to the environment from the household sector are wastes or residuals. They are the unwanted by-products of the consumption activities of households. Similarly, not all of the material inputs that enter the production sector are embodied in the consumption goods flowing on to the household sector. These are the wastes or residuals from production. Thus, there is a flow of residuals from both the production and consumption sectors back to the environment.

These materials flows must obey the basic law of physics governing the conservation of matter. In an economy with no imports or exports, and where there is no net accumulation of stocks (plant, equipment, inventories, consumer durables, or residential buildings), the mass of residuals returned to the natural environment must be equal to the mass of basic fuels, food, minerals, and other raw materials entering the processing and production system, plus gases taken from the atmosphere.[2] This is the principle of materials balance. This principle must hold true for each sector of the economic system taken separately, and for the economic system as a whole. Thus, in the absence of inventory accumulation, the flow of consumer goods from the production sector to the household sector must be matched by an equal mass flow back to the environment.

From this simple model, it is clear that the environment is of considerable value to man as a source of material inputs for production and consumption. However, it has not been as widely recognized—at least until recently—that the environment is also valuable as a receptor for the corresponding residuals flows. Indeed, the environment has an enormous capacity to accept, absorb, and assimilate most of the types of returning materials. But as we shall see below, when this absorptive and assimilative capacity is overused or misused, pollution and environmental degradation is the result.

A more detailed picture of the materials balance model is presented in Figure 2-2 for a developed economy. Here the special

---

[2] Of course, this neglects the conversion of minuscule amounts of matter into energy by nuclear reactors producing electricity.

Figure 2-2   Schematic depiction of materials balance model. *Source.* Reprinted by permission from Allen V. Kneese, Robert U. Ayres, and Ralph C. d'Arge, *Economics and the Environment: A Materials Balance Approach*, Washington: Resources for the Future, Inc., 1970.

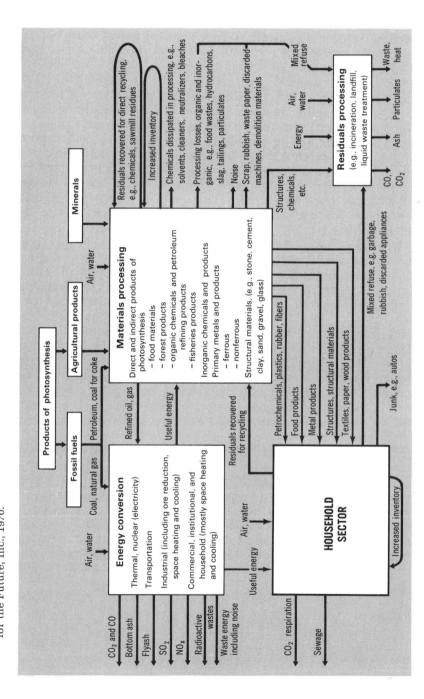

importance of the energy industry as a source of residuals flows is recognized by giving energy conversion a separate box in the diagram. In effect, the production sector of Figure 2-1 has been divided into the energy conversion and materials processing sectors. Inputs into the economy include minerals, the products of recent photosynthesis—agricultural products and timber for construction and paper, fossil fuels which are the preserved products of past photosynthesis, and water. Finally, the atmosphere provides oxygen to support combustion and animal respiration.

The energy conversion sector, through the chemical rearrangement of the material inputs, releases useful energy to the materials processing sector. However, virtually all of the *material* inputs to energy conversion are returned to the environment in the form of gaseous oxides and solid ash. The materials processing sector makes use of air, water, the products of photosynthesis, and useful energy from the energy conversion sector and provides the household sector with a variety of goods and services. As a by-product, this sector also produces a variety of residuals materials including slag, scrap, unrecovered chemicals released to the atmosphere in water, and other residues from raw material refining operations. The residuals from the household sector and its consumption activities include solids (trash), liquids (sewage), and gases (respiratory carbon dioxide and the combustion products of home heating and transportation).

As shown in the diagram, the solid and waterborne wastes of the household and material processing sectors may go through another stage of processing before being returned to the environment. For example, incineration is a way of converting solid residuals into gases. This processing, however, only changes the form and ultimate destination of the residuals flow. The mass of material to be returned to the environment is unchanged.

Figure 2-2 shows flows of energy in addition to materials flows. Energy balances could be drawn up to account for the division of energy production among useful work, noise, and waste heat being dissipated into the air and water. The law of conservation of energy dictates that all energy inputs into the economic system— as well as materials inputs—must eventually find their way back to the environment in some form such as waste heat. In fact, some have suggested that the ultimate limit on economic growth may come not from the scarcity of resources, or overpopulation, but the environment's limited capacity to absorb heat residuals from the economy.[3]

One of the lessons learned from this materials balance model is

[3] See Chapter 3, pp. 40–43.

that the common economic usages of the terms "input," "output," and "consumption" are misleading. For example, the input-output relationships of the production sector are more accurately described as processes of throughput. Similarly, the household sector may use the services of the goods it receives or it may transform these goods by mechanical, chemical, or biological processes; but it never consumes them in a physical sense. After the household sector is through with a good, there is still something left and it must be disposed of, somehow and somewhere. Thus the processes of resource use, production, and consumption (as the economist usually uses that term) can better be described as processes of materials and energy throughput and balanced materials flows. More importantly, these processes are intimately bound together with the problems of residuals disposal—and hence, air, land, and water pollution.

MATERIALS FLOWS IN THE UNITED STATES ECONOMY

Ideally, one would like to have a complete description of the materials flows of the United States economy. Such a description would relate the total quantities of each kind of input used, the sectors of the economy and industries that use the input, how it is transformed back into residuals in production and consumption processes, and the exact quantities of all kinds of residuals "outputs" returned to the environment. While research along these lines is being carried out, only the sketchiest figures on throughput and the United States economy are available presently. Some aggregate figures for the United States economy for the years 1963 to 1965 are presented in Table 2-1.

That table shows that the total weight of materials input to the energy conversion and materials processing sectors was about 2.6 billion tons in 1965.[4] This does not include the oxygen input to combustion or inert construction materials such as sand, gravel, and stone.[5] About 4 to 5 percent of the total input came from net importation of raw and partially processed materials.

[4] The choice of tons as a unit of measurement is for convenience. The reader should be aware that this greatly oversimplifies the situation. The effect of one ton of residuals deposited in the environment depends on its composition as well as the time and place of the discharge. The environmental dangers posed by a ton of DDT are much different from a ton of carbon dioxide.

[5] Since the use of the latter in the economy can be viewed as simply relocation, more like excavation and grading than manufacturing and consumption, these materials have been excluded from the materials balance calculation.

Table 2-1 Weight of Basic Materials Production in the
United States, plus Net Imports, 1963–1965
(Million Tons)

	1963	1964	1965
Agricultural (including fishery wildlife and forest) products			
Food and fiber			
Crops	350	358	364
Livestock and dairy	23	24	23.5
Fishery	2	2	2
Forestry products			
(85% dry weight basis)			
Sawlogs	107	116	120
Pulpwood	53	55	56
Other	41	41	42
Total agricultural	576	596	607.5
Mineral fuels	1337	1399	1448
Other minerals			
Iron ore	204	237	245
Other metal ores	161	171	191
Other nonmetals	125	133	149
Total other minerals	490	541	585
Grand total[a]	2403	2536	2640.5

[a] Excluding construction materials, stone, sand, gravel; and other minerals used for structural purposes, ballast, fillers, and insulation, for instance. Gangue and mine tailings are also excluded from this total. These materials account for enormous tonnages but undergo essentially no chemical change. Hence, their use is more or less tantamount to physically moving them from one location to another. If these were to be included, there is no logical reason to exclude material shifted in highway cut-and-fill operations, harbor dredging, landfill plowing, and even silt moved by rivers. Since a line must be drawn somewhere, we chose to draw it as indicated above.

Source. R. U. Ayres and A. V. Kneese, "Environmental Pollution," in *Federal Programs for the Development of Human Resources,* a Compendium of Papers submitted to the Subcommittee on Economic Progress of the Joint Economic Committee, United States Congress, Vol. 2 (Washington, D.C.: Government Printing Office, 1968). Some revisions have been made in the original table.

In the entire United States economy, accumulation as inventories accounts for about 10 to 15 percent of basic annual input; so that the outflow of residuals back to the environment is correspondingly smaller—about 2.2 to 2.4 billion tons. The accumulation of materials as inventories, however, is not uniform among categories of inputs. Accumulation is primarily in the form of construction materials in the forest products and the other minerals categories. It is likely that almost the entire mineral fuels and food and fiber inputs represent throughput and, hence, result in equivalent amounts of residuals returned to the environment. Perhaps three-fourths of the total weight of throughput is discharged into the atmosphere as carbon and hydrogen in combination with atmospheric oxygen. This results largely from fuel combustion, food "combustion," and incineration as a means of disposal of wastes. The remaining throughput is either in the form of other gases, dry solids such as scrap, trash, and junk, or "wet" solids carried in suspension or solution into rivers or streams. Without adequate statistics, it is difficult to present a more accurate picture of the return flow side of the throughput process. But we lack good data on the outputs of all residuals, particularly atmospheric residuals. Indeed, it is far easier to obtain data on the amount of oil produced in the economy than it is to measure the quantities of carbon dioxide, water, and sulfur oxides, for instance, actually emitted by the combustion of all of this fuel. Much of the air pollutant emission data presented in the next chapter is based on known and fairly reliable measures of fuel production and use rather than independent measures of emissions.

Two items in Table 2-1 deserve further comment. The first is the predominant contribution of mineral fuels to the materials inputs to the economy. By implication, the combustion of these fuels is a major contributor to environmental pollution on the output side. About 55 percent of total throughput (not counting oxygen) is in the form of mineral fuels. The second item is the throughput of pulpwood. The 56 million tons shown for 1965 represent the dry weight of pulp logs delivered to paper mills. About 60 percent of the log by weight emerges from the mill in the form of paper. The remaining 40 percent is residual organic material, most of which is dumped into water courses. Hence, the reputation of the pulp and paper industry as the country's second largest industrial water polluter. And of course, most of the 33.6 millions tons of paper ultimately becomes a residual that must be disposed of.

One indication of the significance of materials throughput for the United States economy can be gained by comparing the physical volume of throughput with the dollar value of the nation's gross national product. GNP for 1965 was about $680 billion. This works out to about eight pounds of throughput for every dollar of GNP. Clearly, however, there is no necessary and fixed relationship between GNP and materials throughput. In fact, it seems inevitable that throughput per dollar of GNP will have to decline substantially if economic growth is to continue along with an

acceptable level of environmental quality. Kenneth Boulding fore-
sees the day when the most accurate view of our world will be as
a space ship, neither taking materials in nor releasing residuals
to the environment.[6] In his view, throughput will be something
to be minimized rather than maximized. How can this be done?

To see how this spaceship earth model can operate, it must be
recalled that residuals do not necessarily have to be discharged
to the environment. In many instances, it is possible to recycle
them back into the productive system. The materials balance view
underlines the fact that the throughput of new materials necessary
to maintain a given level of production and consumption decreases
as the technical efficiency of energy conversion and materials utilization and recycling increases. Similarly, other things being equal,
the longer time that cars, buildings, machinery, and other durables
remain in service, the fewer new materials are required to compensate for loss, wear, and obsolescence.[7] Fully efficient combustion of (desulfurized) fossil fuels would leave only water, ash, and
carbon dioxide as residuals, while nuclear energy conversion need
leave only negligible quantities of material residuals (although
thermal pollution and radiation hazards cannot be dismissed by
any means). These and other throughput reduction possibilities
will be discussed in a later section.

THE ENVIRONMENT AS A RESOURCE

A Definition

The usual dictionary definitions of the word environment men-
tion surroundings, conditions, and influences. A composite of
these definitions would provide us with something like this: the
environment is the totality of natural external conditions and
influences that affect the way things live and develop. Although
this definition is complete, it is unmanageable. In refining it, we
must first put the environment in the economist's usual frame of
reference and then establish some limits on the nature and type
of external conditions and influences that will be included.

As economists, we are interested in the totality of natural
external influences only as they affect man *directly* or *indirectly*.
This does not so much limit the range of environmental conditions

---

[6] Kenneth Boulding, "The Economics of the Spaceship Earth," in Henry
Jarrett, ed., *Environmental Quality in a Growing Economy*, Baltimore: Johns
Hopkins, 1966.

[7] It should be remembered, however, that the use of old or worn machinery
(for example, automobiles) tends to increase other residuals problems.

that are relevant as it establishes the terms on which they are to be evaluated. For example, we are obviously interested in the effect of air pollution on human health. This is a direct effect on man. Are we also interested in the effect of air pollution on the health of Sequoia trees in the Sierra Nevada Mountains? People visit the Sequoia forests and enjoy the experience. Air pollution would indirectly reduce the welfare of those who would experience a loss if the forests were reduced or destroyed by air pollution. In fact, even those who have no present plans to visit the Sequoia forests might be adversely affected if air pollution destroyed their *option* to do so in the future.[8] Furthermore, people who know they will never visit certain natural areas still place a value on preserving or maintaining those areas; and in the absence of better knowledge about ecological systems and the changes man is causing, we cannot be sure that we will not all be affected by some unperceived linkages among different systems. It seems that all of us are likely to have a stake in the continued existence of the Grand Canyon, the redwood forests, and the Alaskan tundra. Thus, although our definition of the environment and our evaluation of environmental changes both make man the measure of all things, they encompass not only short-run and direct effects on man but indirect and long-run effects as well.

The second way of limiting our conception of the broad term, environment, is to view the environment as an asset or a kind of nonreproducible capital good that produces a stream of various services for man. These services are tangible (such as flows of water or minerals), or functional (such as the removal, dispersion, storage, and degradation of wastes or residuals), or intangible (such as a scenic view).

In refining our definition of the environment still further, we will identify several types of services yielded by it that are of value to man and that are affected by the way he organizes and conducts his economic activities. More precisely, these environmental services will be those that are affected by the production and consumption activities of man and by the ways in which he disposes of his residuals. In this way, we can legitimately exclude from our discussion such dimensions of the environment as housing or the political, social, or cultural climates in which man lives.

[8] The concept of "option value" was first discussed by Burton Weisbrod. "Collective Consumption Services of Individual Consumption Goods." *Quarterly Journal of Economics*, 1964. For further discussion of option value and other ways in which man has a stake in preserving the natural environment see John V. Krutilla, "Conservation Reconsidered," *American Economic Review*, 51, No. 4 (September 1967), 777–786.

The Services of the Environment

As the materials balance model showed, the environment performs a valuable service for the economy by *dispensing, storing,* or *assimilating* the residuals generated as a by-product of economic activity. The ability of the environment to serve as a waste receptor stems from the natural processes that transform and/or disperse waste products into harmless areas or harmless and sometimes valuable substances.[9] Wind currents disperse potentially harmful concentrations of air pollutants. Rain and gravity remove pollutants from the air. Indeed, in some cases, air pollutants undergo chemical transformations to less harmful substances. For example, carbon monoxide picks up a second oxygen atom to become carbon dioxide. Bacteria in water feed on and transform the organic wastes into inorganic nutrients for algae, the first link in the aquatic food chain.

If the environment's capacity to absorb or assimilate wastes were unlimited, there would be no pollution problem. Residuals could be dumped into the environment without limit and without cost. Actually, however, the so-called assimilative capacity of the environment is limited in several ways. For example, in the water, bacteria feeding on organic residuals use oxygen. If oxygen supplies are depleted, other forms of aquatic life become adversely affected. Also, the products of this organic decomposition enrich or fertilize the water, and this may result in undesirable changes in the marine ecology. Also, in some cases, the environment has no assimilative capacity at all. Mercury and some other heavy metals are cases in point. In the case of mercury, all of the natural processes appear to work in a perverse manner. Small organisms can convert metallic mercury into organic forms that have adverse health effects. Furthermore, the organic mercury compounds tend to become concentrated in the food chain by natural processes. As a result, a relatively small amount of mercury, which is originally widely dispersed throughout the environment in the nontoxic metallic form, eventually winds up in harmful and in some cases even lethal concentrations in our swordfish and tuna.

These examples plus ordinary observation are enough to establish the proposition that the assimilative capacity of the environment is limited. As a consequence, residuals disposal can impair other environmental functions and services.

A second and most important class of services involves the

[9] For a more complete description of some of these natural processes, see American Chemical Society, *Cleaning Our Environment: The Chemical Basis for Action,* Washington, D.C., 1969.

support of human life. The environment provides a hospitable habitat for man and other forms of life. However, the environment, especially the atmosphere, becomes less hospitable—indeed, habitable—as concentrations of residuals accumulate in it, thereby causing ill health and shortening man's life expectancy. These residuals concentrations can also affect other life forms. There is a growing body of evidence that demonstrates the damaging effects of atmospheric pollution on plant life—both crops and noncommercial species. Moreover, this damage can seriously alter natural ecological systems, causing further harm to man in the long run.

A third class of services provided by the environment could be called *amenity services.* Certain parts of the environment are pleasant spots in which to vacation, spend a weekend, or walk through. The environment can provide amenity services simply by being pleasant. People can utilize these services where they are available by pursuing recreation activities such as hiking, camping, boating, or by purchasing or renting residences in attractive sections of the environment. Just what makes some parts of the environment more attractive to some people or more valuable for some amenity uses is difficult to establish. We know very little about how people perceive their environment or how these perceptions are influenced by differences in cultural values or experiences. Also, tastes and preferences vary among people. Nevertheless, we know, for example, that a lake can be made unfit to swim in or very unpleasant to live next to because of the disposal of human wastes or even septic tank flows into the lake. And we know that living downwind from a Kraft papermill can be an annoying and perhaps unhealthy experience.

Finally, as the materials balance model showed, the environment serves as a source of *materials inputs* to the economy. These include fuels, lumber and minerals, water from rivers, gases from the atmosphere, and fish from the seas. Such materials flows can also be impaired in quality and quantity by residuals discharges, thus raising the cost of obtaining food and materials from the environment.

To summarize, we have defined the environment as a kind of natural asset or nonreproducible capital good which is the source of economically valuable direct and indirect services to man. These services include residuals absorption or waste receptor services as well as life-sustaining, amenity, and materials supply services. These services are all economic goods in the sense that people are willing to pay to receive more of them or to avoid a reduction in the quantity or quality of the services that they provide.

Environmental Quality and Pollution

Given our definition of the environment, we can now attempt a definition of environmental quality. Environmental quality can be defined as the level and composition of the stream of all of the environmental services, *except* the waste-receptor services. In principle, the ultimate measure of environmental quality is the value that people place on these nonwaste-receptor services or their willingness to pay. Since many of the environmental services that comprise environmental quality do not pass through markets —that is, they are not bought or sold—prices or values are not recorded for them. However, the fact that the values are not recorded does not mean that values do not exist. Nor does it imply that people would not be willing to pay money to obtain more of these services, to avoid their loss, or to avoid reductions in their quality because of pollution-induced damages. The fact that prices are not recorded for environmental services and that people cannot, and hence do not, pay for these services reflects the failure of the market system. This will be discussed more fully in Chapter 4.

The willingness to pay for the nonwaste-receptor environmental services constitutes the demand for, or the benefits of, environmental quality. The willingness to pay for the existing level of environmental quality is a measure of the nonmonetary income or welfare accruing to individuals because of the presence of environmental services. This nonmomentary income is no less a part of their real income or welfare than their willingness to pay for the marketable goods and services they consume. Hence, people may be better off as a consequence of a pollution control plan that lowers their money income and consumption but increases the flow of environmental quality services to them.

With these definitions, we are now in a position to define pollution. In economic terms, pollution is the reduction in environmental quality caused by the disposal of residuals. Whenever the disposal of residuals occurs so as to damage life and property or to impair or reduce the quantity or quality of environmental services, pollution exists. With this definition, we can also talk about the costs of pollution. These costs are measured by the value of the nonwaste-receptor environmental services foregone because of the disposal of residuals. If an act of residuals disposal reduces the value of other environmental services by $10, this is the cost of, or the damages of, pollution. Pollution costs are opportunity costs. They can be identified and defined even if they cannot presently be measured precisely or given dollar values.

Notice that pollution is not synonymous with residuals discharge. If it is possible to discharge some quantities of residuals into some parts of the environment so as to cause no discernible or measurable change in the environment, then pollution has not occurred. Even if physical changes can be observed and measured, it may be that they have no impact on the other services that man is deriving from the environment. Again, by our definition, this is not pollution. But we must also be aware of the possibility that wherever residuals discharges occur, it may be the insensitivity of our instruments or our ignorance that prevents us from perceiving the more subtle, long-range, and indirect damages that our actions are causing.

By the definitions given here, environmental quality is maximized when no pollution costs are generated by residuals being returned to the environment. But just as the use of the waste-receptor services has an opportunity cost—for example, foregone amenity services—environmental quality also has an opportunity cost: the foregone waste-receptor services of the environment. As a result, higher environmental quality entails higher costs for managing the residuals flows from the economy. As we pointed out in Chapter 1, and as we will discuss later, this poses the fundamental question of choice concerning management of the environment. In order to maximize the total value of services from the environment, we may have to incur additional waste disposal costs in return for improved environmental quality. A major public policy issue is how much additional cost are we willing to incur in order to control pollution and secure improvements in the quality of the environment.

THE OPTIONS FOR CONTROLLING POLLUTION

Now that we have observed how the throughput associated with production and consumption processes causes pollution in the environment, we can consider the ways in which the production-consumption-pollution relationship can be altered. The materials balance model provides a useful framework for dealing with this set of questions. At this point, we are not concerned with pollution control policies or questions of environmental quality management. Instead, we want to know what the technical alternatives are.

Within the materials balance framework, four basic technological options for pollution control can be identified. All are costly because they require the diversion of resources—land, labor, and capital—from the production of other valuable goods and services. As a result, the undertaking of pollution control activities entails either a smaller basket of goods from the production sector or a basket with a different and less desirable composition. However, these costs must be compared with the benefits of reduced pollution damages and a more desirable environment. Briefly, the four options are: reduce the rate of throughput of materials and energy; treat the residuals to make them less harmful to the environment; choose the time and place of discharge so as to minimize the damage; or augment the waste assimilative capacity of the environment through investment. Each of these options is now considered.

Reducing Throughput—Four Possibilities

If pollution exists, it is a clear signal that the rate of throughput and residuals disposal exceeds the capacity of the environment to absorb the flow without reducing its other nonwaste-receptor services. In such a case, reducing throughput will reduce pollution, other things being equal. Or to put it differently, assuming no changes in the degree of utilization of the other three technological options for controlling pollution, there is a positive relationship between the rate of materials throughput and the level of pollution. More throughput means more pollution, and vice versa. Even if none of the other technological options were used, there are four ways by which pollution can be controlled by reducing the rate of materials throughput in the economy.

The first and most drastic is to *reduce the rate of production* and therefore consumption of all goods and services by the same proportion. This would entail reducing the level of economic activity or GNP, but holding its composition constant. Since each dollar of GNP brought with it eight pounds of materials throughput (in 1965), reducing all throughputs proportionately would cost $250 of GNP per ton of throughput reduced. This technological option is likely to be the most costly of all because it neglects any possibility of altering the ratio of throughput per dollar of final output. Indeed, anything that lowers the materials throughput per dollar of GNP will reduce the cost of obtaining any given level of pollution control. Let us now turn to the ways of altering this throughput-GNP relationship.

The most direct approach for reducing this ratio is to *increase the technical or engineering efficiency* by which the materials inputs are used in the production process, that is, to increase the ratio of usable output to material input. There are numerous possibilities for accomplishing this in almost every industrial operation. For example, in the potato chip industry a more costly method of peeling the potato takes less of the potato with the skin and can reduce the quantity of organic residuals from the plant by up to 50 percent at a given rate of potato input. Conversely, if this type of peeler were installed, the same rate of output could be maintained with a smaller potato input and a smaller residuals outflow. The technically more efficient processes that are available are often not adopted because at existing prices it is cheaper for the businessman to throw away some of the material than to try to make better use of it. The costs to the business of increasing technical efficiency may be greater than the monetary benefits that it obtains. This occurs because the business gains nothing if it reduces pollution—no one pays it to reduce the flow of materials used or residuals returned to the environment.

To better understand why these technically superior processes are often not adopted, consider how the plant engineer decides what level of efficiency in materials use he will adopt. In principle, he could choose any level of efficiency between 0 and 100 percent. At 100 percent efficiency, there would be no residuals at all. Starting at any level of efficiency, for example, 50 percent, an increase in the efficiency of resource use is equivalent to using less of the material input per unit of output. Greater efficiency means less expenditure on material inputs per unit of output. The lower expenditure on inputs is the benefit of increased efficiency. But in order to obtain the higher technical efficiency of materials use, the engineer must use more labor and/or capital in the production process, and this is a cost. The plant engineer will increase the efficiency of resource use whenever the incremental or *marginal benefit* of one more step (say from 50 to 55 percent) is greater than the incremental or *marginal cost* of taking the step.[10] If the step reduces material cost by $10 and it requires only $5 worth of additional labor to accomplish the step, it pays to take it. As efficiency rises, the costs of each additional step get higher and higher, while the additional savings get smaller. We can assume that the plant engineer has already increased the efficiency of resource use to the point where marginal benefits just equal marginal costs, thus exhausting all the possible economic (monetary) gains.

What events could make the plant engineer choose a still more efficient level of utilization? If the price of his material input rises, he will be induced to make more efficient use of the inputs he buys. Also, if some technological change or reduction in the price of labor or capital reduces the cost of greater technical efficiency,

---

[10] In the economists' technical language, marginal cost means the cost associated with a unit increase in production. Similarly, the marginal benefit is the increase in total benefits associated with a small increment in production.

increased efficiency of materials use would become relatively cheaper, and the plant engineer would shift to more efficient techniques. Finally, if it became more costly to dispose of the unutilized residual, it would pay to reduce the amount being wasted by increasing materials efficiency.

A second way of altering the relationship between material inputs and GNP is through recovering residual materials and recycling them as material inputs. For example, some processing chemicals can be recovered and reused. While they were formerly thrown away, they can now be used as inputs for making other products. Paper, metals, and glass can also be recovered after use and recycled as material inputs. In all of these cases, the recycled materials replace new inputs of materials from the environment. Therefore, they simultaneously reduce the flow of residuals to the environment and reduce the flow of new materials from the environment to the economy.

Many opportunities for recycling are now technologically possible, though not economically attractive. The main barrier to recycling and materials recovery is the cost of recovery relative to the value of the recovered materials. If the costs of disposing of residuals into the environment becomes increasingly borne by the disposers, or if the price of new material inputs from the environment rises, or if technological change lowers the cost of materials recovery, increased recycling will become a more economically feasible option.

In some cases, the effects of recycling on residuals flows is more complex than is apparent at first glance. Consider the case of paper. Even in the pre-ecology days of 1965, about 25 percent of each year's production of paper was recycled. There is, no doubt, considerable opportunity to increase the degree of recycling. Each ton of scrap paper collected by a Boy Scout troop or the Salvation Army is only about 60 to 75 percent reusable pulp. The rest of the weight is fillers, coatings, color and ink. Reprocessing the paper involves separating the usable pulp from these other materials. Inherent in this is the necessity for disposing of other materials. At least with the processes presently used, these materials emerge as solid wastes suspended in the stream of water that comes out of the pulping process. This load of suspended solids may be up to ten times greater than the suspended solids generated by making pulp from new pulpwood. Thus, recycling paper involves a combination of reducing one flow of residuals (the pulp component of waste paper) while at the same time simply moving another flow of residuals (that is, the unusable solid portions of the paper) from one place to another.

On balance, this is probably an improvement. But it is a useful example to bring home the point that it is literally impossible to do only one thing where the environment is concerned. Almost every action taken has multiple effects, some of which may be good, and some of which are almost certainly bad. And it takes careful analysis and some expertise to trace out all of the possible effects and ramifications and to arrive at a judgment as to whether the net effect is positive or negative.

Finally, the relationship between GNP and materials throughput can be altered by *changing the composition of GNP*. Some goods involve larger quantities of materials throughput per dollar value than do others. If the composition of output shifts so that goods that generate low residuals increase in production while high residuals-high pollution goods decrease in production, the overall residuals flow can be reduced while the dollar value of GNP remains constant.

There are several possibilities for accomplishing this kind of shift. For example, existing products can be made more durable through design changes and the use of different materials. If they are made more durable, they would need to be replaced less often and, as a consequence, would generate a smaller flow of residuals. Another possibility is to use entirely different technologies to produce a needed service. For example, the air pollution caused by the internal combustion automobiles could be reduced by shifting to electrical cars or by greatly increasing the use of electric powered rail transportation in urban areas. But here again is a case where it is impossible to do only one thing. A shift to an electric transportation system would require an increase in the production of electricity, itself a major source of materials and energy throughput and residuals.

A further possibility for altering the composition of final output is by inducing shifts in consumption patterns. For example, consumers might be induced to shift their demands away from material goods and toward services such as educational and cultural activities. But again the net effect of such a shift is not entirely clear. We must look at both the direct and indirect inputs that go into producing a particular good or service. Services such as education and government create indirect demands for other goods that may well have significant residuals flows associated with their production. Governments need paper. Energy is required for heating, lighting, and transportation. The net environmental impact of such a shift in demand is not immediately obvious.

To summarize, the rate of materials throughput can be reduced by (1) reducing the overall level of economic activity, other things equal, or (2) decreasing the material input requirements for producing a given level of GNP. This reduction in material inputs per dollar of GNP can be accomplished by (a) increasing the technical efficiency of materials use, (b) materials recovery and recycling, and (c) altering the composition of output.

Treatment of Residuals

Residuals treatment is a conventional term that covers a wide variety of possibilities for biologically, chemically, or mechanically changing the form of a waste flow so as to make it less damaging to the environment. The most important point to remember about treatment comes from the materials balance principle. That principle demonstrates that when a residuals flow of one form is reduced, another must be increased. For example, sulfur oxides and particulates or ashes in the emissions of a thermal electric power plant can be removed by various processes; but, they do not disappear. They must either be disposed of as solid or liquid residuals or they must be recycled.

A classic example of a failure to take this fact into account occurred in New York City not too long ago. The city passed regulations aimed at reducing the amount of air pollution caused by the trash incinerators in the apartment houses throughout the city. For most of the apartment houses, substantial expense was required to meet the new standards. In many cases, this expense was more than the apartment house owners were able or willing to incur. Vigorous enforcement of the regulations would have resulted in the shutting down of large numbers of old incinerator systems. When this in fact began to happen, the city found itself with a large increase in the quantities of solid wastes placed by the curbside for pickup. These had to be disposed of through the Sanitation Department. The Sanitation Department was overwhelmed. In the end, the city was unable to cope with the solid wastes to be collected and disposed of, and found it expedient not to enforce the incinerator regulations.

Picking Your Time and Place

Some locations in the environment are better suited for assimilating residuals than are others. Large rivers with rapid flows of water can carry off and dilute larger quantities of wastes than can slower or smaller rivers. An effluent flow, which would turn a small stream into an open sewer, might have almost no perceptible impact on a larger river. Similarly, locations with high winds and good air circulation have a greater capacity for waste disposal than do more stagnant airsheds. Since we are concerned with reducing the impact of residual discharges on the environment, we should take advantage of all possible opportunities to discharge residuals where they will be least noticed and more easily assimilated. Savings in cost and improvements in environmental quality can be significant when a real effort is made to select the proper place for residuals discharges. For example, computer studies of the Delaware River have shown that substantial costs savings can be realized by piping water-borne wastes from sources close to the center of Philadelphia to locations further downstream where the assimilative capacity of the river is greater.

The effect of a residuals discharge also depends on the time of discharge. Timing is important because of fluctuations in the capacity of the environment to assimilate or absorb wastes. Most of us are now familiar with the role of weather conditions that create air pollution episodes. Certain weather patterns impede the flow and mixing of air masses and trap air pollutants in pockets near the source of discharge. At other times, air mixing and flow is quite high and airborne wastes are carried off quite rapidly. Similarly, during times of high temperature and/or low river flow, the assimilative capacity of rivers is impaired. Discharges can be timed to coincide with the periods of maximum assimilative capacity either by halting the residuals producing activity during low capacity periods or by storing the residuals produced until the critical period has passed. Most major cities now have established procedures for dealing with acute episodes of high air pollution on this basis. Many of these plans call for reducing or eliminating automobile traffic in the center city, and/or shutting down center city electric power plants and industrial operations when certain critical air pollution levels are reached.

Investing in Environmental Capacity

The waste assimilation capacity of the environment can be augmented by various kinds of man-made investments. For example, someone once suggested that the smog problem in the Los Angeles basin could be solved by constructing a large set of fans at the foot of the San Gabriel Mountains to blow the smog over the mountains and out of the basin.[11] There are other less startling and more plausible suggestions that have, in fact, been used successfully. One such possibility is to build dams close to

[11] Out of kindness, we have not tried to find this reference and give it a full citation.

the source of rivers with heavy waste loads. The reservoirs are used to store water during periods of high river flow and to release it during periods of low flow when additional water is needed to dilute residuals flows. The release of the reservoir storage in effect flushes out the river. Another possibility is called reoxygenation. A critical factor in the water quality of rivers is the level of oxygen dissolved in the water. Oxygen is consumed during the process of decomposing and assimilating organic wastes. The capacity of the river to assimilate these wastes can be increased by augmenting the natural supplies of dissolved oxygen. This can be done by paddlewheel-type arrangements that agitate the surface of the water, or by diffusers or bubblers that pump oxygen directly into the lower depths of the river.

LESSONS FROM THE MATERIALS BALANCE MODEL

The materials balance model and our definition of the environment as a service-producing resource provide some helpful insights into the environmental problem. They also reveal some general principles that will be very useful in the more detailed discussion of air and water pollution that will follow in subsequent chapters.

Perhaps the most valuable use of the model is in identifying the range of technological options. The variety of ways by which pollution can be controlled is immense, especially when one considers the large number of environmental problems involved. Yet most of the public policy steps taken to control pollution focus on only a narrow range of technological alternatives. For example, the main step taken by the federal government to control water pollution has been to speed the construction of waste water treatment facilities for cities. Relatively little attention has been given to encouraging recycling, improvements in the utilization of resources, or changes in the composition of GNP. In later chapters, as we evaluate alternative pollution control policies, one criterion will be the extent to which they encourage consideration of the widest range of technological alternatives.

A second principle that the materials balance model impresses on us is the interdependency among the various kinds of residuals flows. As long as one always recalls that materials flows must balance, foolish and perhaps costly errors can be avoided when decisions about pollution control policy are made. New York City's failure to see the connection between its regulations concerning incinerators and its solid waste disposal problem has already been mentioned. In another city, scrubbing devices were installed in the smokestacks to remove sulfur dioxide from the exhaust gases of coal-burning electric power plants. But the sulfur-laden waste water from the scrubbers was simply discharged into a nearby river. Reduced air pollution was accompanied by more water pollution. The planners overlooked the fact that materials removed from one residual flow must still be put somewhere and depending on where they are put, they can still cause environmental damages.

The materials balance model can also illuminate the relationships among population growth, economic growth, and pollution. The model says that, other things equal, the more production and consumption, the more resources coming in, and the more waste products going out. But it is not so much the size of the population or its growth as it is what that population is doing—producing and consuming—that is giving rise to our pollution problem. For example, in the period since 1940, 90 percent of the increase in production and consumption of electrical energy and its associated throughput can be attributed to the growth in *per capita* *demand* for energy; while only 10 percent can be attributed to *population growth* in the United States.[12] In other words, if over the last thirty years our population had remained stable, electrical energy production would only be about 10 percent below its present levels. If economic growth continues without any change in the composition of GNP, materials throughput and hence pollution will continue to grow. But as the discussion of technological options should make clear, it is not economic growth per se that is causing the growth in pollution, but the form that economic growth has taken.

Finally, the materials balance model also sheds light on the problems of establishing public institutions for environmental management. It suggests that the following principles should be adhered to in creating new public agencies. The environmental-quality management agency should exercise comprehensive authority over air, water, and land pollution so that it can recognize and accommodate the interdependencies among residuals flows. For example, the decision as to whether to transform an airborne residuals flow into a waterborne residuals flow should be made by an agency responsible for both air and water quality. The agency's jurisdiction should extend over the major physical systems involved, for example, a river basin or an air shed. In its planning it must be able to treat the system as a unit. Only in this way can it fully utilize the option of augmenting assimilative

---

[12] See Hans Landsberg, "A Disposable Feast," *Resources*, June 1970.

capacity and fully realize any economics of scale in treating wastes.[13]

In summary, the comprehensive view of the environment and the economy provided by the materials balance model warns us against taking an overly narrow and partial view of the pollution problem. Air, water, and land pollution cannot be considered separate, unrelated problems. Nor can the environment be managed in isolation from the economic sectors making use of it. The income levels, tastes and preferences, technologies, and prices that determine the outcome of the economic system have corollary effects on materials throughput and the environment. But the effects go both ways. Environmental-management policies will affect the costs and prices of different goods in different ways, leading to changes in the nature and composition of economic output. And, in turn, these changes will themselves affect the nature of the residuals flows and their impact on the environment.

MANAGING THE ENVIRONMENT

Having defined the environment as an asset-like stock yielding flows of services to businesses and households, and having used the materials balance model to show how the environment interacts with the economy, we can now discuss how the services of the environment can be allocated—managed—so as to maximize social welfare. In a very real sense, environmental services are no different from the services of other scarce resources—labor, land, and capital. And like the other scarce resources, we must make choices about allocating environmental resources among competing uses. However, unlike most other kinds of resources, choices involving the allocation of environmental resources cannot be left to individuals acting separately in unregulated and decentralized markets. Alternative institutions and mechanisms must be created through public action for the purpose of managing environmental services and guiding them to their best use.

One of the uses to which scarce environmental resources can be put is as waste receptors. Our concern with the pollution problem

[13] But herein lies a fundamental problem of institutional reform. Political boundaries seldom correspond to the natural boundaries of air and water systems. Yet, the public agency's authority should not be weakened by divided political jurisdictions within its boundaries. Hence, these principles for guiding institutional development constitute a counsel of perfection. But they do point out the direction that institutional reforms must take.

See A. Myrick Freeman III and Robert H. Haveman, "Water Pollution Control, River Basin Authorities, and Economic Incentives: Some Current Policy Issues," *Public Policy,* 19, No. 1 (Winter 1971), 53–74.

is evidence that we have exceeded the normal waste assimilation capacity of the environment. This is a predictable consequence of the fact that the waste receptor services of the environment have no prices attached to them and are available free of charge to anyone who wishes to use them. All that is needed is a pipe and ditch, or a tailpipe, or a smokestack. Yet, the use of the environment for waste disposal generates social costs in that other environmental services are impaired. Hence, there is a divergence between the price that the user of the environment for waste disposal must pay and the costs that are imposed on society because of this use. Because the environment is not effectively owned by anyone, it is controlled by no one. As a result, it is overused and abused. In a smoothly functioning market economy, this overuse and abuse of resources is normally avoided by the device of ownership and the charging of prices. Prices allocate and ration scarce resources to their highest use. However, there are no markets for waste disposal services of the environment. Hence, the misallocation and waste of environmental services.[14]

Although pollution causes damages by impairing other environmental services, the converse is also true: the control of pollution reduces these damages and enhances environmental quality. But this enhancement also has its cost. As a result, basic management questions are posed: "How far should we go toward improving the environment?" "How much pollution control or environmental quality do we want to buy?"

The logic of economics suggests the following kind of answer. We should take a number of small steps in the direction of improving the environment. With each step, we should examine the benefits or values associated with the improvements in environmental quality and compare these with the costs of controlling the pollution that led to the improvement. As long as the benefits from the step exceed the costs that the step has imposed, we should move ahead one more step. We would expect the benefits of further improvements in the environment to be smaller as environmental quality rises; and the costs of additional improvements should rise as we take on more and more pollution control. Eventually, a point will be reached at which the incremental or marginal benefits are just equal to the incremental or marginal costs. A further step means that we are buying environmental improvement that costs more than it is worth to us—and this is wasteful. We should choose that level of environmental quality at which the marginal benefits equal the marginal costs.

---

[14] The causes and results of the failure of markets to efficiently allocate environmental resources will be discussed in Chapter 4.

The process we have just described is just another way of saying that there is a demand curve (marginal benefits) and a supply curve (marginal costs) for environmental improvement. And, if we only knew what the curves looked like and acted accordingly we would move to the point where the demand and supply curves intersect. Such a diagram is shown in Figure 2-3. At environmental quality level Q_1, the marginal cost of a small improvement is $3, and the marginal benefits are $5. Environmental quality should be improved to Q_2 where the marginal benefits just balance the marginal costs. At Q_3 a small improvement costs $2 more than it is worth.

There are two additional points to be brought out about environmental management. The first is that it is important that we utilize the least costly or most efficient combinations of technological options in obtaining improvements in environmental quality. This is because the amount of environmental improvement that society will ultimately decide to buy depends on its price or cost. The higher the costs of environmental improvement in terms of

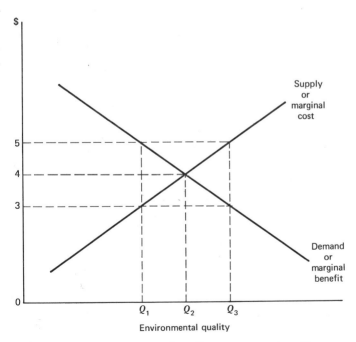

Figure 2-3 The environmental quality management problem

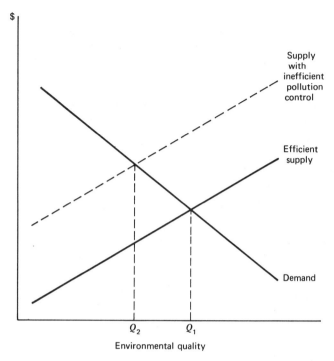

Figure 2-4   Environmental quality and inefficient pollution control

reduced consumption of goods and services, the less environmental quality we will buy. In Figure 2-4, the dashed supply curve shows the effect of using inefficient pollution control options. The costs of environmental quality are higher, and the amount of environmental quality that would be chosen is smaller. It is shown as Q_2.

The second point concerns the likely trend of environmental quality over time. It is likely that, as incomes rise over time, the demand curve for environmental quality will shift out to the right. In Figure 2-5 this is shown by the shift to D'. If the supply curve does not shift, this analysis suggests that we will devote some of our increased affluence to buying more environmental quality. But an unchanging supply curve is unlikely. Rising GNP and materials throughput will make any level of environmental quality more costly to obtain. Hence, over time this force would tend to shift the marginal cost or supply curve upward and to the

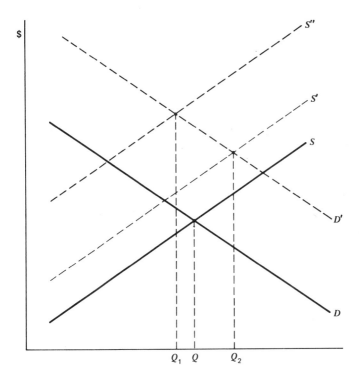

Figure 2-5 Environmental quality over time

left—from say S to S". If this were the only supply (or cost) side
effect of economic growth, the optimal level of environmental
quality would decrease from Q to Q_1. However, this effect is likely
to be partially offset by improvements in the technology of pollution control that accompany economic growth. These make it
possible to obtain a given level of environmental quality more
cheaply, other things equal. Such technological improvements in
controlling pollution tend to shift the supply curve down to the
right—say from S" to S'. As we have depicted in Figure 2-5, the
effect of economic growth and increased affluence indicates an
increase over time in the optimal level of environmental quality
from Q to Q_2.

Whether environmental quality will actually improve or degrade
over time will depend on two things. First, will we be able to

develop and implement public policies to get us on the efficient supply curve of Figure 2-4 and out to the hypothetical intersection of the demand and supply curve? Assuming we are successful in devising such policies, the second question is whether the supply curve will shift upward and to the left more rapidly than the demand curve shifts out and to the right over time. We will return to this theme in Chapter 8.

3

Some Facts About Environmental Pollution

This discussion of the physical and technical aspects of pollution problems begins by concentrating on those problems, or potential problems, that affect the entire planet.

These "global" problems pertain largely to the atmosphere. The marks of man are and have been clearly seen on that entire thin film of life-sustaining substance. We will discuss other large-scale problems, or potential problems, particularly those related to the "biosphere," only briefly. After focusing on these global problems, we will discuss "regional" problems. By regional we mean problems other than global. It is necessary to use a word like regional rather than terms pertaining to political jurisdictions such as nations, states, or cities. This is because the scale of pollution resulting from the emission of materials and energy follows the patterns, pulses, and rhythms of meteorological and hydrological systems rather than the boundaries of political systems. In discussing some of the regional aspects of pollution, we will concentrate on the three primary environmental media—land, water and air. Our discussion, therefore, will focus on water pollution, air pollution, and the solid wastes problem.

GLOBAL PROBLEMS

It seems to have come as somewhat of a shock to the natural science community that man not only can, *but has,* changed the composition of the whole atmosphere.[1] It is essential that we find

---

[1] The fullest discussion of the range of problems considered in this section will be found in *Man's Impact on the Global Environment: Assessment and Recommendations for Action,* Cambridge: M.I.T. Press, 1970; and *Inadvertent Climate Modification: Report of the Study of Man's Impact on Climate,* Cambridge: M.I.T. Press, 1971.

out more not only about the direction and magnitude of these changes, but how and why they have taken place.

Before proceeding to what may be real global problems, let us dispose of one red herring. One of the specters raised by the more alarmist school of ecologists is that man will deplete the world's oxygen supply by converting it into carbon dioxide in the process of burning fossil fuels (coal, oil) for energy. Indeed, it has been suggested that the arrival of this catastrophe might be hastened as man's pollution interrupts or impedes the photosynthetic replenishment of oxygen by trees and other green land and marine plants. This idea has now been thoroughly discredited by two separate pieces of evidence. The first comes from measurements of the changes in the level of oxygen in the atmosphere over a period of years. These observations have shown the oxygen content of the atmosphere to be remarkably stable.[2] The other piece of evidence—perhaps more persuasive—is in the form of an experimental calculation. If one were to burn, hypothetically, the entire known world supply of fossil fuels and all the present plant biomass, the calculated impact on the oxygen supply would be to reduce it by about 3 percent. This is much too small to be noticed in most areas of the earth.

Potentially significant and serious effects on the atmosphere and climate are more likely to be connected with changes in carbon dioxide and particulate matter (including aerosols) in the atmosphere, the injection of waste energy into the atmosphere, and petroleum in the oceans. We will discuss each of these briefly in turn.

The production of carbon dioxide (CO_2) is an inevitable result of combustion of fossil fuels. In contrast to oxygen, the relative quantity of CO_2 in the atmosphere has increased measurably. The rate of increase of the volume of CO_2 appears to have been about 0.2 percent per year.[3] Although earlier, and more alarming estimates put the possible increase in CO_2 at about 50 percent by the end of the century,[4] more recent studies support a predicted increase of only between 15 to 20 percent.[5]

The difference in estimates is accounted for by the newly recognized fact that less of the CO_2 generated by combustion is staying in the atmosphere than was previously supposed. Apparently one or more of the "sinks" for CO_2 is responding to the

[2] L. Machta and E. Hughes, "Atmospheric Oxygen in 1967 to 1970," *Science,* 168, No. 3939 (June 26, 1970), 1582–1584.

[3] *Man's Impact on the Global Environment,* op. cit., p. 47.

[4] Conservation Foundation, *Implications of Rising Carbon Dioxide Content of the Atmosphere,* New York, 1963.

[5] *Inadvertent Climate Modification,* op. cit., pp. 235–238.

increased concentration, or possibly even a third force is leading to greater absorption. The main sinks for CO_2, or more specifically carbon, are solution in the oceans and conversion by the flora of the earth. Perhaps the availability of carbon dioxide is the limiting factor to growth for some of these plant populations, and they are responding to its increased availability from the atmosphere. Another possibility is less reassuring. Somewhat anomalously the mean temperature of the earth's surface has been falling over the past couple of decades according to Weather Bureau observations. As is true of many gases, the solubility of CO_2 in water increases when water temperature falls. Also, higher concentrations of CO_2 in the atmosphere could cause some increase in absorption by the oceans, since CO_2 solubility in water depends in part on its concentration in the atmosphere above.

The cumulative effects of an increase in CO_2 may be significant for the earth's climate. Carbon dioxide absorbs infrared radiation. Most of the incoming solar radiation is in the form of visible light. But the earth's temperature balance is maintained by the reradiation of excess energy largely in the form of infrared radiation. Therefore a buildup of carbon dioxide in the atmosphere would tend to reduce the reradiation of energy and cause the surface of the earth to rise in temperature.

An increase in CO_2 of 15 to 20 percent might raise the air temperature near the surface of the earth by about 0.5°C. This is a small enough change relative to other forces at work so that predictions of the effect of this change are very uncertain. However, if the rate of increase were to continue for another 100 years (to 2100), the world's mean surface temperature could be raised as much as 2.0°C. This likely would be accompanied by melting of ice caps, inundation of seacoast cities, and undesirable temperature increases in densely inhabited areas.[6]

While increases in carbon dioxide levels might tend to heat the planet, the second possible effect of man's activity on world climate is in the opposite direction. Some meteorologists believe that man's industrial and agricultural activities are causing the world to cool off.[7] The suspected mechanism is an increase in particulates which, they think, are increasing the earth's albedo (ability to reflect incoming solar radiation). Farming and other activities in arid areas and the combustion of fuels send immense amounts of particulates into the atmosphere each day to act as an umbrella or

[6] Ibid., pp. 239, 128–129.
[7] R. A. Bryson and J. T. Peterson, "Atmospheric Aerosols: Increased Concentrations During the Last Decade," *Science, 162*, No. 3849 (October 1968), pp. 120–121.

shade from the sun. This is an undisputed fact. What is in dispute is the likely long-term effect of this man-generated increase in albedo.

Some not only believe this effect is significant, but that it may be sending us rather rapidly toward an ice age—perhaps the final ice age resulting in a perpetually frozen planet. Other factors might lead in the same direction, as we shall see subsequently. The freeze-up hypothesis is, however, disputed by other meteorologists who argue that it is important to recognize the difference between particulates of different types and at different elevations in the atmosphere. One meteorologist has pointed out that there has been a large amount of volcanic activity in recent years that has deposited great quantities of particulates at high elevations in the atmosphere.[8] The net effect of these is fairly clear—to reflect more energy away from the earth. This could well be responsible for the observed decline in the earth's temperature. On the other hand, he points out that particulates deposited at relatively low altitudes, such as those generated by man, could well have the reverse effect because they reflect energy back toward earth as well as away from it. Calculations tend to show that the former effect has so far outweighed the latter. Thus, when the effect of the volcanic particulates wears off over a period of years, the lower altitude particulates could begin to reinforce CO_2 as a factor leading to rising world temperatures.

A third global factor possibly affecting earth temperatures is the release of energy to the atmosphere due to the energy conversion activities of man. A large proportion of the energy from the fuels man uses is transferred directly to the atmosphere—for example, the energy converted in automobile engines. Another large proportion is initially transferred to water—for example, when condensers in electric power plants are cooled with water. But this too is rather quickly injected into the atmosphere by induced evaporation in watercourses or cooling towers. Thus essentially all of the energy converted from fuels is sooner or later transferred to the atmosphere as heat. Because this is true, it is possible to make a rather precise estimate of this transfer by calculating the energy value of the fuels used in the world. On this basis there is an average emission of about 5.7×10^{12} watts of energy from energy conversion by humans.[9]

[8] J. M. Mitchell, Jr., "A Preliminary Evaluation of Atmospheric Pollution as a Cause of the Global Temperature Fluctuation of the Past Century," S. Fred Singer, ed., *Global Effects of Environmental Pollution*, Holland: Reidel, 1970, pp. 97–112.

[9] W. R. Frisken, "Extended Industrial Revolution and Climate Change," unpublished report, Resources for the Future, Inc., Washington, D.C., 1970.

What does one make of such a monstrous number? For comparison this is equal to about 1/15,000 of the solar radiation absorbed by the earth's crust and the atmosphere. That doesn't seem like much. However, another important element in the picture is the fact that energy conversion is a rapidly growing activity all over the world. The most spectacular example is the conversion of fossil fuels to electric power which in the United States has been proceeding at a doubling time of ten years and even faster in one or two other large economies. Worldwide energy conversion as a whole has been proceeding at a growth rate of about 4 percent a year. If this rate of growth is continued for 130 years, energy rejected to the atmosphere will have grown to about 1 percent of the absorbed solar radiation. This could be enough to have a substantial effect on world climate as indeed it is now contributing to the higher temperatures of cities relative to outlying areas. If growth continues at 4 percent for another 120 years, the emission of waste energy into the atmosphere will have reached 100 percent of the absorbed solar radiation. This would be a total disaster. The resulting mean increase in world temperature would be about 50°C—a condition totally unsuitable for human habitation. We will never reach such a situation, but the important question is what circumstances will prevent us from doing so.

If one is given to apocalyptic visions, he can readily imagine a situation in which CO_2, particulates, and energy conversion reinforce each other and, after a short reprieve from the volcanoes, transform the earth into a kind of mini-hell. But other things may happen too. For one thing, we are annually spilling on the order of 1.5×10^6 tons of oil directly into the oceans with perhaps another 4×10^6 tons being delivered by terrestrial streams. This may be enough "oil on troubled waters," some scientists believe, to smooth the sea surface sufficiently to cause its reflectivity to be increased significantly. Again, the increased associated albedo would tend to cause cooling. But at the same time the reduction in the atmosphere-ocean interface would tend to diminish the absorption of CO_2 and thus possibly raising CO_2 levels and leading to a warming condition.

There are other potential global threats, too, for example, the supersonic transport (SST). The European Concorde and the Russian SST have already flown, and the United States version may yet get off the ground one day. Aside from the major question of sonic booms, the emissions from SSTs may have substantial effects on the upper atmosphere. SSTs would fly at 65,000 to 70,000 feet where the atmosphere is quite different from that at sea level. It is extremely dry, and air layers at that elevation do not mix readily with the lower atmosphere. Five hundred SSTs might be in operation by the mid-1980s. If these were the American type, their emissions might cause an increase of water vapor in the upper atmosphere of 10 percent globally and possibly 60 percent over the North Atlantic where most of the flights would occur.[10] This could give rise to large-scale formations of very persistent cirrus clouds. Furthermore, the emissions of soot, hydrocarbons, nitrogen oxides, and sulfur oxides will result in the formation of other kinds of particles as well. The effects of all this would be somewhat uncertain, but presumably not unlike those produced by particulates deposited into the upper atmosphere by volcanoes—in other words, increased albedo and consequent cooling at the earth's surface.

More recently another possible route to ecological catastrophe through the SST has been uncovered. The SST will inject oxides of nitrogen into the stratosphere. These oxides may react chemically with a stratospheric layer containing a high concentration of ozone. This ozone layer presently shields the surface of the earth by absorbing harsh ultraviolet radiation from the sun. If this ozone shield were significantly reduced in size, life on the surface of the earth would be exposed to ultraviolet radiation in substantial amounts and with unpredictable but undoubtedly adverse consequences. Recent calculations have suggested that SST flights at the proposed frequency and altitude may reduce this ozone shield by a factor of 2.[11]

A final category of substances of possibly global significance is persistent toxins such as pesticides, mercury, and other heavy metals. For example, DDT has been found in living creatures all over the world. How it got to remote places like the Antarctic is still somewhat mysterious, but apparently substantial amounts are transmitted through the atmosphere as well as through the oceans. DDT and other pesticides may trigger large-scale changes in ecological systems, with some species becoming extinct. Possibly these persistent toxins could affect the O_2-CO_2 balance by poisoning the phytoplankton (algae and other minute aquatic plants) which are involved in one of the important CO_2-O_2 conversion processes. Persistent pesticides could also have a serious adverse effect on the productivity of estuarine and offshore fisheries which are important food sources for some countries. The long-term consequence of these and other kinds of major ecological changes are not well understood.

[10] Man's Impact on the Global Environment, op. cit., pp. 100–107.

[11] Harold Johnston, "Reduction of Stratospheric Ozone by Nitrogen Oxide Catalysts from Supersonic Transport Exhaust," Science, 173, No. 3996 (August 6, 1971), 517–522.

Clearly, we are operating in a context of great uncertainty. Man has the potential through his activities now and in the relatively near-term future, to affect the world's climatic and biological regimes in a substantial way. It seems beyond question that a serious effort to understand man's effects on the planet and to monitor those effects is essential. If it turns out that worldwide controls on such things as the production of energy and CO_2 are needed, we will face an economic and political resource allocation problem of unprecedented difficulty and complexity.

AIRBORNE RESIDUALS

The discussion of global effects of pollution was necessarily somewhat speculative. We now examine problems of a less grand nature. The regional problems related to air, water, and land pollution are easier to study and measure, and hence are *comparatively* better understood. Since many of the analytical points in this book are illustrated by reference to particular cases, some familiarity with the physical and technical aspects of pollution is desirable. The aim of the remainder of this chapter is to give the reader some familiarity with the terminology used to describe air and water pollution, its effects, and means of abating pollution.[12] In addition, we attempt to summarize what is now known about the present level and trends in air and water pollution in the United States.

Air Pollutants and Their Effects

There is virtually an infinity of airborne residuals that may be discharged to the atmosphere, but the ones of central interest and most commonly measured are carbon monoxide, sulfur dioxide, oxides of nitrogen, hydrocarbons, and particulates. The direct and observable effects of these pollutants on people range in severity from the lethal to the merely annoying. Except for extreme air pollution episodes, fatalities are not, as a rule, directly traceable individually to the impact of air pollution. Instead, air pollution is an environmental stress that, in conjunction with a number of other environmental stresses, tends to increase the incidence and seriousness of a variety of diseases, including lung cancer, emphy-

---

[12] For a more thorough and comprehensive discussion of these topics see American Chemical Society, *Cleaning Our Environment: The Chemical Basis for Action*, Washington, D.C., 1969.

sema, tuberculosis, pneumonia, bronchitis, asthma, and even the common cold. Acute air pollution episodes, that is, short periods of high air pollution, have been correlated with higher death rates and rates of hospital admissions. Such occurrences have been observed in Belgium, England, Mexico, and the United States, among others. But the more important health effects appear to be associated with persistent, chronic (that is, relatively low levels for long periods of time) exposure to the degraded air that exists in most large cities.[13] Because each of the airborne residuals discussed here can be linked with some adverse effect on people and/or property, their presence detracts from the services people gain from the environment and environmental quality depends on the *absence* of these substances.

In terms of weight, *carbon monoxide* is the most important air pollutant. Most of the carbon monoxide comes from the incomplete combustion of gasoline in the internal combustion automobile engine. The primary short-term effects of carbon monoxide are well-known—impairment of mental and physical functions, and at high concentrations, death. The symptoms of dizziness, headache, and lassitude occur at concentrations that are not infrequently recorded at street levels in our cities. Little is known about long-term effects of repeated exposures.

The coal and oil that are burned for space heating and electric power generation contain elemental sulfur as an impurity. When the fuel is burned the sulfur also burns, producing *sulfur dioxide* and, to a much smaller extent, *sulfur trioxide*. When placed into the atmosphere the latter substance is immediately converted into sulfuric acid. For humans, acute (higher level and short duration) exposure is known to have adverse effects on the functioning of the lungs. Sulfur oxides can also cause damage to vegetation, exterior paints, and other materials. There is now a substantial body of evidence that chronic (low levels but long time periods) exposure has serious adverse health effects and is associated with higher mortality rates.[14]

Several kinds of *oxides* of *nitrogen* are formed in combustion processes in automotive transportation and electric power gen-

---

[13] See Lester B. Lave, and Eugene P. Seskin, "Air Pollution and Human Health," *Science, 169*, No. 3947 (August 21, 1970), 723–733; Lester B. Lave, "Air Pollution Damage: Some Difficulties in Estimating the Value of Abatement," in Allen V. Kneese, and Blair T. Bower, eds., *Environment Quality Analysis: Theory & Method in the Social Sciences,* Baltimore: The Johns Hopkins Press, 1972; and William Hickey, "Air Pollution" in William Murdock, ed., *Environment: Resources, Pollution, and Society,* Stamford: Sinauer, 1972.

[14] Ibid.

eration. The oxides of nitrogen are primarily of concern because of their contribution to the formation of photochemical smog, a phenomenon that we discuss below. In addition, nitrogen dioxide has been found to be harmful to the lungs, has had adverse health effects on laboratory animals, and has caused leaf damage and growth reduction to plants. And there is increasing evidence of higher mortality rates being associated with chronic exposures to nitrogen oxides.[15]

Unburned and partially burned *hydrocarbons* are emitted primarily through the exhaust pipes of automobiles. Miscellaneous sources such as evaporation of gasoline and other volatile hydrocarbons in gas stations, dry cleaners, gas tanks, carburetors, and even drying paints also contribute substantial quantities. Hydrocarbons at observed atmospheric concentrations have no known directly harmful effects. However, a number of hydrocarbons at low concentrations react photochemically in the atmosphere with the nitrogen oxides to produce smog.

The fifth major class of pollutants consists of a heterogeneous mixture of suspended solids and liquids known as *particulates*. Specific particulate substances may be directly harmful to humans, for example lead from automotive exhausts. Others may cause serious animal and vegetable damage, for example fluorides. The most common forms of particulates are dust and ash from combustion. Chronic exposures to particulates have been associated with higher death rates and other health effects.[16] They also may have synergistic relationships with other air pollutants.[17]

Photochemical smog is not a residual. Instead, it is a mixture of gases and particles manufactured by the sun chiefly out of the nitrogen oxides and unburned hydrocarbons emitted from the combustion of gasoline. The major components of smog are oxidizing by nature, and as a class they are called *oxidants*. Eye irritation, although possibly not the most important effect of the oxidants, is the most troublesome and most commonly recognized effect in urban atmosphere. At commonly observed levels oxidants make it more difficult for people to breathe, especially people already suffering from respiratory disease. Also important are the effects of oxidants on plant life. Serious damage has been known to occur to shrubs and leafy vegetable and forage crops.

In addition to the five major pollutants, there is a substantial number of chemicals and metallic compounds that are released in

---

[15] Hickey, "Air Pollution," op. cit.
[16] Lave and Seskin, op. cit., and Lave, op. cit.
[17] Synergistic refers to an interaction between two substances in which the total effect is greater than the sum of the parts.

relatively small quantities from various mining and manufacturing processes. Some of these, for example, fluorides from phosphate fertilizer plants, may have quite severe but localized effects. There is another substance that is causing increasing concern because of accumulating evidence that the severity of the problem has been underestimated and, because it is so widespread. The substance is lead, and a major source is the automobile. Several studies have shown that people with high occupational (toll collectors, traffic policemen) and residential (adjacent to freeways) exposures have alarmingly high amounts of lead in their body. And inner-city children may be in a most dangerous situation because of the combined effects of lead ingested from the atmosphere and from lead paints and dusts in slum housing.

Sources, Levels, and Trends

The five primary air pollutants are listed in Table 3-1, as well as the activity emitting them into the atmosphere. The table shows that almost 300 million tons of pollutants were spewed into the air in 1969. Over one-half of that total is accounted for by the transportation industry, primarily the automobile. Other studies have shown that the automobile accounts for about 60 percent of the air pollution in smaller cities and up to 90 percent in some of the large cities. The automobile is the largest single source of hydrocarbons and nitrogen oxides, which combine to form smog. Fuel combustion, primarily oil and coal at electric power generating plants, is the major source of sulfur oxides. Coal-burning power plants are also major sources of particulates.

Although Table 3-1 focuses on the number of tons of pollutants, this information by itself is somewhat misleading. It does not, after all, give any indication of the relative effects of the different pollutants on, say, human health. For example, data on the aggregate weight of pollutants disguises the fact that one ton of sulfur oxides is of far more danger to human health than is a ton of carbon monoxide.

These data also hide two other important characteristics of air pollution—variation in the severity of the air pollution problem over time and from one part of the country to another. In any one well-defined geographic region, say a city, the quality of the air in that region at any point in time will depend on several conditions. Among them are: the rate at which pollutants are being discharged in that region; the air temperature; wind conditions; humidity; cloud cover; and other weather conditions. All of these factors tend to change markedly over time. For some pollutants,

Table 3-1 Emissions of Air Pollutants by Source, 1969 (in Millions of Tons per Year)

Source	Carbon Monoxide	Particulates	Sulfur Oxides	Hydro-carbons	Nitrogen Oxides	Total	Percent Change 1968–1969
Transportation	111.5	0.8	1.1	19.8	11.2	144.4	− 1.0
Fuel combustion in stationary sources	1.8	7.2	24.4	.9	10.0	44.3	+ 2.5
Industrial processes	12.0	14.4	7.5	5.5	.2	39.6	+ 7.3
Solid waste disposal	7.9	1.4	.2	2.0	.4	11.9	− 1.0
Miscellaneous	18.2	11.4	.2	9.2	2.0	41.0	+18.5
Total	151.4	35.2	33.4	37.4	23.8	281.2	+ 3.2
Percent change 1968–1969	+ 1.3	+10.7	+5.7	+1.1	+4.8	+3.2	

Source. President's Council on Environmental Quality, *Environmental Quality—1971*, Washington, D.C., 1971, p. 212.

such as smog, there are well-defined daily cycles that can be traced to peaks in auto traffic (which govern emission levels) and the noonday sun. These cycles are less pronounced or absent on weekends (no commuter traffic) or on cloudy or windy days. For other pollutants, cycles are not pronounced and concentrations tend to vary randomly as does the weather. This variability over time in air pollution levels must be taken into account in trying to assess the effects of air pollution and in planning air pollution control programs.

If all of the pollutants dumped into the atmosphere were spread evenly over the entire United States, we would not have nearly as serious an air pollution problem as we now have. But as population has tended to concentrate itself in major urban areas, pollution also exhibits geographical concentration. This is illustrated in Table 3-2. Note carefully the change in dimension

Table 3-2   Pollution Concentration by Population Size, Annual Averages (Micrograms per Cubic Meter)

Population	Total Suspended Particulates	Sulfur Dioxide	Nitrogen Oxides
Nonurban	25	10	33
Less than 10,000	57	35	116
10,000–25,000	81	18	64
25,000–50,000	87	14	63
50,000–100,000	118	29	127
100,000–400,000	95	26	114
400,000–700,000	100	28	127
700,000 to 1 million	101	29	146
1 to 3 million	134	69	163
More than 3 million	120	85	153

Source. President's Council on Environmental Quality, *Environmental Quality—1971*, Washington, D.C., 1971, p. 243.

from tons of pollutants discharged to concentrations in the atmosphere (for example, micrograms per cubic meter in the air). Also, these data represent an average of conditions over a full one-year period. The table shows the average concentrations of three important pollutants for cities of various sizes. These data demonstrate that the air pollution problem is much more serious in large metropolitan areas than it is in small towns and nonurban places. For example, the concentration of suspended particulates is over 5 times greater in cities containing 1 million people than it is in nonurban sites; sulfur dioxides concentrations are over 8 times as great in very large cities as they are in nonurban places. As striking as the absolute tonnage of pollutants and their geographical concentration is their growth over time. If one concentrates only on the total tonnage of pollutants, the trend over time is dramatically up. For example, while electric utilities released about 3 million tons of sulfur oxide into the air in 1940, by 1970 they were spewing out nearly 20 million tons. That annual increases in the total weight of pollutants are continuing is shown in Table 3-1. Emissions of all five of the pollutants in 1969 exceed emission levels in 1968. Overall, the total increase in pollutants by weight is 3.2 percent from 1968 to 1969. On the other hand if trends in emissions by source are examined, it appears that emission controls on new autos and tougher regulations by some cities on trash incinerators may have begun to have some beneficial effects.

Again these data hide possible significant changes in pollution levels in some geographic regions. Indeed, while the aggregate volume of pollutants emitted has risen considerably, there is growing evidence that the pollution levels for *some* pollutants in *some* of our largest cities have been decreasing in recent years. Table 3-3 presents evidence on the concentration of two important air pollutants in four major cities. This data seems to indicate that sulfur oxide pollution has been declining in three of the four cities, Chicago showing no clear trend. On the other hand, oxides of

Table 3-3 Air Pollutant Concentrations in 4 Cities, 1962–1968 Annual Averages (in Micrograms per Cubic Meter)

City and Pollutant	1962	1964	1966	1968
Chicago				
Sulfur dioxide	282	458	215	307
Nitrogen oxides	81	87	105	90
Cincinnati				
Sulfur dioxide	92	100	81	44
Nitrogen oxides	56	60	68	60
St. Louis				
Sulfur dioxide	NA	155	113	73
Nitrogen oxides	NA	62	64	43
Washington, D.C.				
Sulfur dioxide	144	126	115	97
Nitrogen oxides	56	68	66	88

Source. President's Council on Environmental Quality, *Environmental Quality—1971*, Washington, D.C., 1971, p. 216.

nitrogen have been about stationary or increasing as in Washington. Some of the reductions are no doubt due to the effective application of emission controls in some of the larger cities. But probably more important is the movement of heavy industry from the central city to suburban locations and the substitution of cleaner burning petroleum products for coal as fuel.

Air Pollution Control

In this section we briefly describe some of the better known techniques for controlling some of the air pollutants discussed above. Opportunities for augmenting the assimilative capacity of the atmosphere are extremely limited. The one device often used is the high stack that releases the pollutants where the wind speeds are higher and there can be greater dispersion and dilution. A related control option that comes under the heading of "picking your time and place" is to regulate the location of new factories that are potential pollution sources. For example, wise land-use planning would bar potential polluters from locating in the middle of or upwind of major population centers or in river valleys where pollutants might tend to get trapped by high surrounding hills.

There are several possibilities for control under the heading of treatment. Suspended particulate emissions from flour milling, cement factories, and other industrial operations can be controlled by enclosing the operations in sheds or buildings and passing the ventilating air through filtering systems before releasing it. There are also processes for removing particulates and sulfur oxides from the exhaust gas streams of power plants and incinerators. The particulates (essentially a very fine ash) which are recovered have sometimes been found to have economic value as construction material. The recovered sulfur oxides can be used to manufacture sulfuric acid or other sulfur products. Unfortunately, however, the sulfur oxide removal processes are not yet well-developed and have not been widely adopted, for reasons that we will discuss in Chapter 7.

The most widely used technique for controlling sulfur oxide emissions involves a reduction of throughput of sulfur. If fuels of low sulfur content can be found, they can be substituted for high sulfur fuels, with a corresponding reduction in sulfur oxide emissions. Unfortunately, natural supplies of low sulfur coal and oil are limited. Some oils are now being processed to remove the sulfur; but sulfur removal from coal is not economically feasible at present prices.

The major source of carbon monoxide, nitrogen oxides, and hydrocarbons is the internal combustion engine used in today's auto. These emissions result from some peculiarities associated with the burning of gasoline in these engines. Broadly speaking, the technological options entail: process changes such as redesign of the engine, changing powerplants (steam car or electric car), or shifting to mass transit; treatment of exhaust gases to transform the pollutants after combustion; or placing restrictions on the entry of cars into the most congested and polluted areas. The special problem of the automobile and some policy alternatives will be discussed in Chapter 7.

WATERBORNE RESIDUALS

Degradable Residuals

It will be helpful in understanding what happens when residuals are discharged to water bodies if we make a somewhat over-simplified distinction between *degradable* and *nondegradable* materials. The most common form of degradable residual is the variety of organic materials ranging from human sewerage to pulp mill wastes. When an effluent bearing a substantial load of degradable organic residuals is discharged into an otherwise "clean" stream, stream biota, primarily bacteria, feeds on the wastes and decomposes them into their inorganic constituents— nitrogen, phosphorus, and carbon, which are basic plant nutrients. As part of this process, some of the oxygen that is naturally present in dissolved form is used up by the bacteria. But this depletion tends to be offset by the reoxygenation that occurs through the air-water interface and also as a consequence of photosynthesis by the plants in the water. If the waste load is not too heavy, *dissolved oxygen* (DO) in the stream first will drop, to a limited extent, and then rise again.

One measure of the quantity of degradable organic materials present in water is the amount of oxygen required to support the complete decomposition of the material. This gives rise to *bio-chemical oxygen demand* (BOD) as a common unit of measurement for organic materials of a widely different composition. BOD measures the number of pounds of oxygen required to decompose the organic matter contained in the discharge.

If the BOD of an organic residual discharge to a stream becomes great enough, the process of degradation may exhaust the dissolved oxygen. In such cases, degradation is still carried forward but it takes place anaerobically, that is, through the

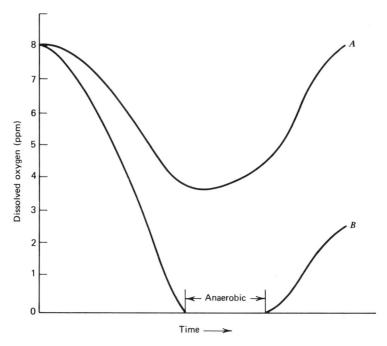

Figure 3-1   The oxygen sag

action of bacteria that use organically or inorganically bound oxygen rather than free oxygen. Gaseous by-products result, among them carbon dioxide, methane, and hydrogen sulfide. Water in which wastes are being degraded anaerobically emits foul odors, looks black and bubbly, and is altogether offensive aesthetically. Anaerobic conditions are common summer occurrences in some United States rivers.

This emphasis on organic residuals and BOD as a measure of their magnitude stems from the essential role of dissolved oxygen in the ecology of rivers and streams. Dissolved oxygen is one of the most important indicators of the life-sustaining capacity of a water body. High levels of DO (7 to 8 parts per million) are essential for the spawning of several important fish species. Most species of fish can manage the rest of the year with lower levels, say 5 ppm. But as DO levels fall to 4 ppm and below, only carp and other trash fish are likely to survive. Low (or zero) dissolved oxygen is the most common cause of fish kills.

The behavior of DO levels as a consequence of the introduction of a quantity of organic matter can be described by a characteristically shaped "oxygen-sag" curve. In Figure 3-1, the oxygen-sag curves show the changes in DO over time for a "one shot" discharge of organic matter. At first DO falls as the bacteria use up the available supply. The rate of decomposition falls and reoxygenation increases, the curve reaches a minimum, and then recovers, eventually to the original DO level. Curve A shows a case in which the assimilative capacity of the water is not greatly taxed. Curve B represents a larger discharge, and shows that decomposition uses up all the available oxygen supply, resulting in a period of anaerobic conditions.

Figure 3-2 shows the profile of dissolved oxygen along a river below a source of continuous organic discharges. Since the flow of the river carries the organic matter downstream as it is being decomposed, the minimum point of the DO sag may occur ten to twenty miles from the source of the pollution.

The precise shape of the oxygen-sag curve depends on a multiplicity of factors including the size of the discharge, stream flow, water temperature, the composition of the discharge, the season, and the wind. Even with a constant rate of discharge, the curve will vary in shape as the other factors change over time. Figure 3-3 shows the high, low, and average oxygen-sag curves for a hypothetical river. Variation in stream conditions over time leads to marked variation in the effects on the river of a given residuals discharge; and this variation makes planning for pollution control more complex. For example, it is irrelevant that the minimum point of the average DO curve is safely above 5 ppm, since the low conditions resulted in DO levels below 2 ppm with fish kills as a probable consequence.

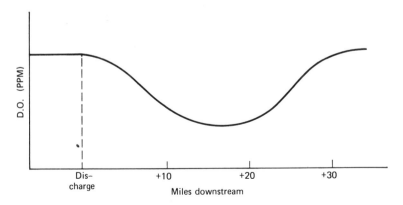

Figure 3-2 The dissolved oxygen profile of a river

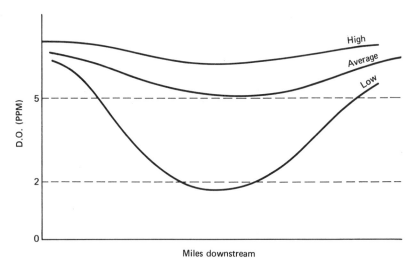

Figure 3-3   The dissolved oxygen profile varies over time

In addition to reducing DO, organic discharges can have other potentially adverse effects on receiving water. When organic material in a stream is decomposed, the materials balance principle still applies. The mass is conserved; only its form is changed. The organic materials are broken down into their inorganic constituents, principally carbon dioxide and water but also including significant amounts of inorganic nitrate and phosphate compounds. These compounds are the nutrients for algae and can stimulate their growth. Up to a certain level, algae growth in a stream is not harmful and may even increase fish food, but large growths can be toxic to fish, produce odors, reduce the river's aesthetic appeal, and increase water supply treatment problems. Large algae blooms themselves create a substantial oxygen demand when they die and are decomposed.

Problems of this kind are particularly important in comparatively quiet waters such as lakes and tidal estuaries. In recent years many lakes have changed their character radically because of the buildup of plant nutrients. The most widely known example is Lake Erie, although the normal "eutrophication" or aging process has been accelerated in many other lakes. Even small relatively undeveloped lakes in Maine and Wisconsin, for example, are experiencing accelerated eutrophication caused by seepage of nutrient laden "treated" effluent from septic tanks. The possibility of excessive algae growth is one of the difficult problems in planning for pollution control—especially in lakes, bays, and estuaries. What are the principal sources of organic wastes? The federal government has compiled estimates of the quantities of organic matter being *generated* by various industrial sectors and by human wastes entering sewers—all measured in BOD. These are presented in Table 3-4. Not all of these wastes are actually discharged,

Table 3-4 Estimated Waste Loads Before Treatment in the United States—1968 (in Millions of Pounds per Year)

Manufacturing	Biochemical Oxygen Demand
Food and kindred products	4,600
Textile mill products	1,100
Paper and allied products	7,800
Chemical and allied products	14,200
Petroleum and coal	550
Rubber and plastics	60
Primary metals	550
Transportation equipment	160
All other manufacturing	470
Machinery	180
Total manufacturing	29,670
Domestic:	
Served by sewers	8,500
Total	38,170

Source. Environmental Protection Agency, *Cost of Clean Water,* Volume II, March 1971.

however. For the nation as a whole about two-thirds of the BOD coming from the domestic sector—households—is removed by treatment before discharge into watercourses. Reliable estimates of the level of treatment of industrial wastes to remove BOD are not available. But it seems likely that the average percentage removal is less than is the case for domestic wastes. Thus it appears that industrial sources are responsible for between three and six times as much BOD discharge as domestic sources. And the three major industrial sources, chemicals, paper, and food products, may each be responsible for as much organic pollution as the domestic sector. Some industrial plants are fantastic producers of degradable organic residuals. In fact one paper mill in Maine discharges almost twice as much organic wastes into the nearby river than is produced in the form of sewerage by the total human population of the state.

There are two other types of degradable pollutants. The first is waste heat, particularly that which is carried by the water used to cool large nuclear power plants. The artificial injection of heat into a river reduces the capacity of the water to hold dissolved oxygen, accelerates oxygen using decomposition, and may have more direct adverse effects on a wide variety of temperature sensitive organisms.

Infectious bacteria might also be included among the degradable pollutants since their source is human sewerage and they tend to die off in watercourses. Because of the universal treatment of drinking water supplies, the traditional scourges of polluted water —typhoid, paratyphoid, dysentery, gastroenteritis—have become almost unknown in advanced countries. But infectious bacteria and waterborne viruses such as hepatitis still pose a health hazard to swimmers and can contaminate seafoods.

Nondegradable Residuals

Once discharged, nondegradable residuals are not decomposed by stream biota. For most of these residuals, the only significant changes that occur are dilution and perhaps settling. This group includes a variety of inorganic chemicals, salts, colloidal matter, and even silt. When these substances are present in fairly large quantities, they can result in toxicity, cloudiness (turbidity), unpleasant taste, hardness and, especially when chlorides are present, in corrosion. They may also cause changes in the stream ecology.

For some nondegradable substances, primarily the salts of some heavy metals, natural processes work perversely. For example, simple mercury compounds are absorbed by the lower forms of plant life, and then, via the food chain, are concentrated to potentially poisonous levels by higher order species. Two particularly vicious instances of poisoning by heavy metals have stirred the population of Japan. These are mercury poisoning through eating contaminated fish (Minimata disease) and cadmium poisoning through eating contaminated rice (Itai Itai disease). Several hundred people have been affected and more than a hundred have died. The Canadian government has forbidden the consumption of fish from both Lake Erie and Lake St. Clair because of feared mercury poisoning, and mercury has been discovered in unsafe quantities in tuna and swordfish marketed in the United States.

Persistent Pollutants

There is a third group of pollutants, mostly of relatively recent origin, that does not fit comfortably into either the degradable or nondegradable categories. These "persistent" or "exotic" pollutants are best exemplified by the synthetic organic chemicals produced in profusion by our modern chemical industry. They enter watercourses as effluents from industrial operations and also as waste residuals from many household and agricultural uses. These substances are termed "persistent" because stream biota cannot effectively attack their complex molecular chains. Some degradation does take place, but usually so slowly that the persistent residuals travel long stream distances, and in groundwater, in virtually unchanged form. Pesticides (for instance, DDT) and phenols (resulting from the distillation of petroleum and coal products) are among the most common of these pollutants.

Some of the persistent synthetic organics, like phenols, present primarily aesthetic problems. The phenols can cause an unpleasant taste in drinking water, especially when it has been treated with chlorine to kill bacteria. Others are suspected of causing public health problems and are associated with periodic fish kills in streams. Some of the chemical insecticides are unbelievably toxic. The material endrin, which until recently was commonly used as an insecticide and rodenticide, is toxic to fish in minute concentrations. It has been calculated, for example, that 0.005 of a pound of endrin in three acres of water one foot deep is acutely toxic to fish.

Concentrations of the persistent organic substances have seldom if ever risen to high enough levels in public water supplies to present an *acute* danger to public health. The public health problem centers around the possible *chronic* effects of prolonged exposure to very low concentrations. However, the existence of chronic effects is extremely difficult to establish conclusively. Similarly, even in concentrations too low to be acutely poisonous to fish, these pollutants may have profound effects on the stream ecology, especially through biological magnification in the food chain; higher creatures of other kinds—especially birds of prey—are now being seriously affected because persistent pesticides have entered their food chains.

The long-lived radioactive materials might also be included in the category of persistent pollutants. They are subject to degradation, but at very low rates. Nuclear power plants are likely to be an increasingly important source of such pollutants. Generation of power by nuclear fission produces radioactive waste products that are contained in the fuel rods of reactors. Periodically, these fuels are separated by chemical processes to recover plutonium or to prevent waste products from "poisoning" the reactor and reducing its efficiency. Such atomic waste can cause substantial environmental damage unless it is disposed of safely.

Currently, a large volume of low-level waste resulting from the day-to-day operation of reactors can be diluted and discharged into streams, although the permissible standards for such discharge have recently been severely questioned in the United States, both outside and inside the Atomic Energy Commission.

Controlling Water Pollution

All four of the technological options described in Chapter 2 have applications for controlling water pollution. The most common is treatment to remove or transform the residual material. *Primary treatment* describes a process of filtering and settling that can remove up to 35 to 40 percent of the BOD from human sewerage, and various percentages of settlable materials from other types of wastes. Since primary treatment entails the physical separation of solids and liquids, there remains the need to dispose of the solid or sludge material. Sludge disposal is a major and growing problem. Organic sludges such as those obtained from primary treatment of human wastes can be incinerated, used for landfill, or processed and used as fertilizer or soil conditioner. New York City and some New Jersey communities transport the sludge by barge to dumping grounds off the New Jersey and Long Island shore, but this is having disastrous effects on the marine ecology. Even the ocean has limits to its capacity for residuals absorption.

Secondary treatment describes several biological processes for treating the organic wastes that remain after primary treatment. These processes are quite similar to those that go on in natural water bodies, differing principally in their speed. A combination of primary and secondary treatment for domestic wastes can remove between 85 and 95 percent of the BOD. Secondary treatment does not result in any by-products such as sludge. However, the end products of the decomposition, principally nitrates and phosphates, remain in solution in the effluent where they may subsequently give rise to nutrient pollution. After primary or secondary treatment the effluent stream can be chlorinated wherever bacterial pollution is a potential problem.

Tertiary treatment refers to a variety of advanced waste treatment techniques. Some of them are designed to increase the percentage of BOD removal, perhaps to as high as 99 percent.

Tertiary treatment may be required in some areas to remove plant nutrients to prevent algae problems and accelerated eutrophication. The capacity of the water body to absorb organic wastes can be augmented either by increasing the rate at which oxygen supplies are replenished (reaeration) or by releasing water stored in reservoirs to increase stream flow and dilution during critical periods. Also, the time and place of discharge can be varied to spread out the impact of effluent discharges over time and along a given stream. For example, two plants on opposite banks of a stream might cause severe pollution problems if their wastes were dumped into the stream at the same point. But if the wastes from one plant were piped downstream beyond the minimum point of the dissolved oxygen sag, the combined discharges could be within the assimilative capacity of the stream. Also, since stream flow is a critical factor in determining the impact of residuals discharge on water quality, withholding or at least reducing discharges during periods of low flow may be valuable. Discharges can be reduced either by storing effluents in lagoons or ponds, reducing plant output, or shutting the plant down. Plant managers can and sometimes do schedule plant overhaul and maintenance periods during the time of expected low flow.

Finally, there is a variety of ways in which residuals flows can be reduced through changes in technical and industrial processes, materials recovery, and recycling. It is difficult to generalize about these possibilities. However, one example may prove useful. The first major step in converting logs to paper is to "cook" the pulpwood to separate the cellulose fibers used in making paper from the lignins and resins that bind the fibers together in the tree. In the sulfite pulping process, after the fibers are separated, the remaining lignins and processing chemicals are usually discharged into nearby streams. A number of papermills have now shifted to the Kraft or sulfate process, which permits a substantial reduction in waterborne residuals. In the Kraft process after the fibers have been separated, the remaining liquor is concentrated through evaporation and then burned. The concentrated liquor is so rich in organic material that its combustion actually provides heat for part of the remaining pulping process. In addition, this evaporation-incineration process permits the recovery and reuse of the chemicals used to digest the pulp log. It should be noted, however, in accordance with the materials balance principle, that the shift from the sulfite to the Kraft pulping processes is not an unmixed blessing. A by-product of the combustion process is a flow of combustion gases containing sulfur oxides.

Nonpoint Sources of Pollution

The above discussion of pollutants, their sources, and the control technologies all relates to what is known as *point source* pollution. By that we mean there is a readily identifiable pipe or ditch through which the residual is transported to the receiving water body. But as these point sources yield to effective pollution control policies, we will find that the nonpoint sources of pollution described here will loom larger in both relative and absolute importance. Indeed, in some rivers it is even now becoming clear that controlling point sources of pollution will have little or no effect on actual water quality.

Agriculture, natural resource exploitation, and nature herself are the major contributors to nonpoint source pollution. Surface and near-surface run-off from cropland, pastures, and animal feedlots are major sources of organic matter, plant nutrients, persistent pesticides, and silt. Similarly, timber harvesting, especially clearcutting, can contribute substantial organic and inorganic waste loads to streams within the watershed. Coal mining exposes sulfur-containing coal and slag to groundwater seepage and weathering, with the result that surface waters become polluted with sulfuric acid. And, since land development means the destruction of natural cover, heavy rains move more quickly and directly into the rivers, carrying with them natural materials which in the quantities involved must be considered to be pollutants.

We will have very little to say about this set of problems in the remaining chapters. This is primarily because the range of public policy strategies that are appropriate to point source pollution does not appear to have much value in dealing with major nonpoint source problems. But the importance of nonpoint sources of water pollution reinforces the case for regional river basin management of water quality. Agencies with such a scope could influence land use to reduce the adverse effects of discharges from nonpoint sources and implement the instream measures for water quality improvement we have discussed previously. For some nonpoint sources the latter may be the most important technological tools available.

SOLID RESIDUALS

Just about every type of object made and used by man can and does eventually become a solid residual. In the United States, about five pounds per person per day of solid wastes are collected, of which about three pounds come from households and commercial establishments. Industrial, demolition, and agricultural wastes constitute most of the remainder. Altogether the United States generates (exclusive of agricultural wastes) each year approximately 200 million tons of solid residuals from household, commercial, and industrial activities and spends about $5.7 billion to handle and dispose of them.[18] In addition, there is a large amount of uncollected solid wastes that litter the countryside.

The disposal of solid wastes can have a number of deleterious effects on society. Littering, dumps, and landfills produce visual disamenities. The disposal of solid wastes can cause adverse effects on air and water quality. Incineration of solid wastes is an obvious source of air pollution in many areas, as are burning dumps. Furthermore, drainage from disposal sites can pollute streams and ground water supplies. The sites may also provide a habitat for rodents and disease-carrying insects. The disposal of collected solid wastes (which excludes much industrial waste, automobiles, and all agricultural wastes) in the United States is roughly in the following proportions: about 90 percent goes into open dump and landfill operations; another 8 percent is incinerated; and a small amount, about 2 percent, goes into hog-feeding and miscellaneous categories. A properly operated landfill can be an effective and low-cost way of disposing of wastes with a minimum of environmental pollution. However, many landfills are poorly operated and impose environmental damages through effects on the air, water, and landscapes.

Automobiles are a special problem. Of the 10-20 million junk cars in existence at any one time in the United States, about three-fourths are in the hands of wreckers and most of the rest are abandoned and littering the countryside. Recovery of scrap materials could be made much more economical by slight design changes, but presently automobile manufacturers have no incentive to make these changes.

CONCLUSION

Having considered the nature of the environment, the effects of residuals on the environment and its users, and the technological means for mitigating the adverse effects of residuals on the environment, we turn now to the economic side of the problem. In Chapter 4 we discuss the nature of market economies and how they can be expected to perform where environmental resources are involved. This is a prelude to considerations of the economic questions of choice—how much environmental quality do we want, and what strategies should we choose to manage our scarce environmental resources.

---

[18] President's Council on Environmental Quality, *Environmental Quality—1971*, Washington, D.C., 1971, pp. 111, 153.

4

The Market System and Pollution

Managing the environment can be viewed as a problem of allocating the services of scarce environmental resources among competing ends or uses. A river can be used for waste disposal or to sustain a salmon fishery of high commercial and recreation value, but it cannot be used for both purposes. The choices are mutually exclusive, and using the river for one purpose has a cost, the foregone opportunity for the other use.

The necessity for choice and the opportunity costs of choices characterize all resources in the economic sphere, not just environmental resources. A man can make shoes or program a computer, but not both at the same time. Forested land can be used to provide lumber for strip mining or as a park, and any one use precludes the others. The tools of economic analysis are designed to aid our understanding of how economic resources—human resources, capital, and land—are allocated among competing uses in a market economy. It should not be surprising that these same economic tools can help to explain how environmental resources are allocated or misallocated in such a market economy. An understanding of how the market system fails to allocate these resources efficiently is essential if we are to offer useful advice on how to make things better.

Economic theory teaches us that under certain conditions markets can solve the problem of resource allocation in an efficient or optimal manner. The ideally functioning market system does this by automatically generating information and signals and conveying them to economic decision makers. These signals are the prices of goods and resources. This information describes the relative gains and costs to a decision maker of using resources of any kind. Through such descriptions, prices assist the decision maker in determining whether more or less of any good or service should be bought or sold. They assist him in determining how the good or service should be allocated among competing uses. Prices, markets, and scarcity also provide incentives for getting the most out of any given bundle of resources or a dollar of expenditure. Where all goods and resources pass through markets that are competitive and where the other necessary conditions are satisfied, prices serve to guide resources and goods into their most beneficial use.

If there are no markets for some valuable resources, goods, or services, or if the markets do not function properly, the resulting prices do not convey the correct signals and the decisions based on those prices are not optimal. *Market failure* has occurred, in that the operation of the market system has not led resources to be allocated to their best uses. Such market failure occurs on a massive scale where environmental resources are concerned. Since environmental services do not pass through markets, markets fail to attach prices to them, and, hence, fail to guide their allocation to the highest value uses. As a first approximation, we can say that pollution is caused by market failure.

The remainder of this chapter is devoted to a more thorough consideration of this basic point. Following this, we turn to an examination of the ways in which people can act collectively through governments to correct for this failure of decentralized markets. If the market system as an institution has failed in allocating environmental resources, the institution must be redesigned or new institutions created to carry out this function. The discussion of public policy alternatives for controlling pollution and managing the environment can perhaps best be viewed as an evaluation of alternative institutional arrangements.

THE MARKET SYSTEM [1]

The term "market system" takes on two meanings as it is used by economists. On the one hand, it refers to an actual functioning apparatus and set of institutional and cultural arrangements that are serving to guide the allocation of resources through the placing of prices on them. On the other hand, the term also is used by economists to refer to an intellectual idealization or "model" of this system and the functions it performs. This model has been developed and analyzed by generations of economists who were curious as to how such an apparently uncoordinated set of activ-

---

[1] For a more extensive introduction to this subject, see Robert H. Haveman and Kenyon A. Knopf, *The Market System*, Second Ed., New York: John Wiley and Sons, 1970.

ities as are observed in economies organized around markets seemed to achieve a rather impressive degree of economic order. Millions of bottles of milk arrive in New York City each day and are distributed to stores and left in front of the right apartment doors. Men and women working for thousands of different employers operate stamping machines, weld metal assemblies, solder wires, and feed IBM cards to a computer. And the result is a Boeing 747. Simply understanding how the market system achieves order out of such a myriad of seemingly unrelated activities is important enough. But economists have also wondered whether this order had certain desirable social properties, that is, whether it in some sense serves the interests of the people. Let us first discuss the functions of any economic system and how a market system performs them, and then return to the question of whether it helps society to maximize its welfare.

Every economic system—whether planned, free market, or traditional—provides a framework for making choices. The necessity for making these choices stems from the fact of scarcity, that is, the limited or finite endowment of resources. The choices to be made are: *What* goods and services are to be produced? What techniques are to be used in making them (*How* will they be produced)? And *Who* will enjoy the use of society's production, that is, for whom are the goods being produced. The need for choice is clear. No economic system yet known to man can produce everything for everybody, using any old technique.

The choice of what kind of economic system shall guide society's allocation of resources is in large part a choice about what group in society is to make the three choices identified here. The market system for resource allocation and economic choice is based on two value premises: that the personal wants of the individuals in the society should guide the use of resources in production, distribution, and exchange; and that the individuals themselves are the best judge of their wants and preferences and best able to act accordingly.

The idealized or model market system that has been so carefully analyzed by economists has the following properties.

1. All markets are competitive. This means that no particular firm or individual can affect any market price by decreasing or increasing the supply of goods and services offered. All firms and individuals are price takers rather than price makers—there exists no market power.

2. All participants in the market are fully informed as to the qualitative characteristics of goods and services and the terms

of exchange among all commodities and services. These terms are called prices.

3. All decision makers in the system are motivated by self-interest and economic gain, and make decisions concerning resource use in accord with those objectives. Resource owners try to maximize the return, both pecuniary and nonpecuniary, that they receive from selling the services of what they own; firms maximize their profits; and consumers allocate their income among the available array of goods and services so as to maximize their satisfaction or well-offness.

4. All resources and goods are individually owned, and the individuals can control these resources. As a result, they can shift these resources freely from one use to another and capture all the benefits of ownership. To state it differently, this condition requires that there be no spillovers or externalities. A spillover occurs when one individual's production or consumption of a good affects another individual either beneficially or adversely.

An economic system possessing these characteristics produces an allocation of resources that can be called "economically efficient." This means that no other allocation that one can imagine would make at least one person better off without making someone else worse off. That is, no other allocation could make everyone better off simultaneously. The actual allocation of resources that comes out of such a system depends on certain other characteristics of the economy. These include

1. The quantity and quality of resources that the society possesses.

2. The technological sophistication by which these are combined in production.

3. The tastes and preferences of consumers.

4. The distribution of ownership of resources that determines the distribution of income and purchasing power among individuals.

Firms play a key role in the functioning of the market system through their purchasing of resources, combining them in production, and selling the goods and services to consumers. Motivated by the prospect of profits, firms strive to purchase resources at the lowest possible prices, to combine them in the most efficient manner, to produce those things with the highest possible value relative to cost, and to sell them at the highest possible price.

The firms that are most successful at this are rewarded by profit. The firms that are unsuccessful due to lack of foresight, technical skill, managerial talent, or luck lose money and eventually fall by the wayside.

Resource owners play an important role by their desire to sell their resources—the services of their labor, their land or their capital—to the highest bidder. This is how resource owners receive income. The firms that are willing to pay the most for the resources must be able to place them in the highest value use. Consumers' tastes and preferences influence resource allocation through their dollar votes. Firms want to gain the revenue generated when people spend their incomes; thus, they are given incentives to produce the things that people want. Consumers' search for the lowest price for given products keeps continuous downward pressure on price. This search provides the competitive discipline among sellers of goods and furnishes them the incentive to minimize the costs of production.

The opportunity for unlimited exchange among individuals, along with the condition that consumers have information about all exchange opportunities and are motivated to seek out opportunities for exchange that will make them better off, combine to result in exhausting all possibilities for mutually beneficial trade. No further opportunities for mutually beneficial exchange can exist (that is, where at least one of the two parties would be better off after the trade), because by assumption all such opportunities would be recognized and taken advantage of. The competition among producers means that there can be no opportunities for increasing profits by changing production techniques or producing more or less of a product, for if any such opportunities existed, they, too, would have been exploited according to the assumptions of the model.

The logic of this analysis can be made more explicit with a supply and demand curve diagram. In Figure 4-1, the demand curve labeled DD shows how much of this good or service individuals would be willing to buy at different prices. It reflects the underlying tastes and preferences of individuals and the existing distribution of income. It can also be interpreted as showing the value buyers place on a unit of the good at the margin, that is, the marginal value or marginal willingness to pay for the good. For example, at a price of P_1, the demand curve shows that individuals are willing to buy Q_1 and no more. If at Q_1 any individual were willing to pay more than P_1 for an additional unit, he would gain by purchasing more, and the total quantity demanded would be greater than that shown by the demand curve. Similarly, none of the individuals who are purchasing the good could have a mar-

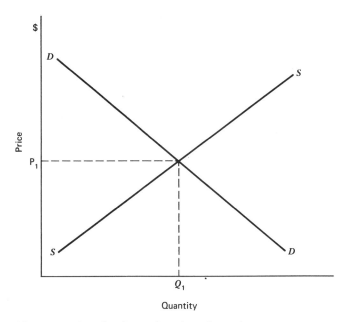

Figure 4-1 Supply, demand, and market price

ginal willingness to pay which is less than P_1, or they would not be purchasing this much of the good.

The supply curve SS shows the amounts that suppliers are willing to sell at different prices. If the market is competitive, this curve also shows the marginal cost of production at each output level, since a producer would not expand production beyond, say, Q_1 unless the higher price covered the increase in his total costs.

If the four properties of a competitive market system are present, there will be supply and demand curves such as this for all goods and services. In each market the price and quantity will be that given by the intersection of the supply and demand curve. At this price, individuals' marginal willingness to pay will just equal the marginal cost of production. This, in fact, is the condition for economically efficient allocation of resources. Consider a case where the marginal willingness to pay is $10 and the marginal cost is $5. This means that if an additional unit of the good is produced, it uses up $5 worth of resources to produce $10 worth of goods, a social gain of $5. Only when such an opportunity exists is it possible to make one individual better off without making someone else worse off. Production of the good should be expanded as long as there is an opportunity to realize such a gain. The market forces of a competitive economy will cause this expansion to occur.

Although we can say that an ideal competitive market system results in an economically efficient allocation of resources, we must recognize two important limitations to such a statement. The first concerns the rather narrow criterion of desirability that is implicit in the statement. The second limitation stems from differences between the real world economic system and the idealized model described here.

The definition of economic efficiency used here focuses on the size and composition of the output generated by the economic system. It is silent on the question of equity or the distribution of this output. In a competitive market economy, the question "Production for whom?" is answered by "Whoever owns the resources," for it is these individuals who receive the income from production and can use the income for purchasing the output. If the ownership of land and capital were divided so that everyone had equal shares, and all persons had approximately equal labor skills, the distribution of income would be approximately equal. In contrast, if most of the people had only low skill levels, and the ownership of capital and land were concentrated in the hands of a few, the resulting distribution of income would be quite unequal. All the preceding analysis of the competitive market system can tell us is that in each case the resulting allocation of resources will be the best that can be gotten *given the initial distribution of resources.*

Because of this, one further assumption must be made if we are to attribute some desirability to the outcome of the market process. We must assume that there is some political mechanism through which collective choices can be effectively made and that through this political mechanism the society has collectively determined the rules that govern the distribution of income and wealth in the society. If these conditions hold, it can be said that the competitive market system yields the best possible allocation of resources. It achieves this optimum because it does the best that can be done with a given technology, resource endowment, set of tastes and preferences, *and* that particular distribution of resource ownership and its associated distribution of income that society has chosen through an effective political mechanism.

The second set of limitations on this analysis stems from the rather unrealistic characteristic of the model. Referring to the four necessary properties of the model market system described previously, we know that not all markets are perfectly competitive. There are monopolies and industries controlled by a few large corporate giants, which have a substantial degree of control over the prices of their products. Labor unions, large financial institutions, and the concentration of land ownership mean that at least some sellers of resources have some degree of monopoly power and can control or at least influence the prices received for their resources.

Moreover, information is imperfect and costly. No one has perfect information on all the alternatives before him. In particular, consumers are never fully informed, and seldom adequately informed concerning the qualitative characteristics of the goods they are thinking of purchasing. Also, we know that tastes and preferences are not strictly given but are subject to the influence of advertising, itself a use of resources.

Not all economic decision makers in firms are motivated strictly by the profit of the corporation. To the extent that the management of a widely held corporation makes decisions that increase its own income at the expense of corporate profits (for example, by increasing salaries and fringe benefits), or department heads in large bureaucratic corporate organizations make choices based on furthering their own careers or building their own empires, market forces are not guiding resources to their highest valued uses.

From the point of view of environmental management, the most important difference between the characteristics of the idealized model and the real world concerns the ownership and control of resources and the presence of spillovers or externalities. In the next two sections, we discuss the role of property rights and ownership in a market economy. We show how the inability to define and enforce private property rights in environmental resources has resulted in large-scale market failure and misallocation of resources. The environment represents an important instance where the outcome of market processes, while perhaps orderly, cannot necessarily be said to have desirable social properties.

MARKET FAILURES

There are two major sources of market failures which are relevant to the problem of environmental pollution. The first is the lack of a well-defined and enforceable system of private property rights in many of the environmental resources. Because no one owns environmental resources, no prices are attached to them. As a consequence, private economic decision makers are not receiving the correct signals concerning the use of these resources. They are not given the proper economic incentives to place these *commonly held resources* in their highest valued use. The second is the public good nature of many environmental services. For reasons to be described in the following paragraphs, public goods require public or governmental intervention in order to ensure that they are produced in appropriate quantities. Private markets will fail to allocate sufficient resources to the production of such goods, resulting in a misallocation of resources.[2]

Property Rights and Market Failure

For markets to function properly, the ownership of a good or a resource must be clearly defined and enforceable. With full ownership, the owner can prevent others from using, benefiting from, or damaging the good without making compensation. When such uncompensated benefits or damages occur, we call them spillover effects or *externalities*. Much of the body of law in western nations is based on the necessity for defining the rights of ownership, protecting them, providing for their transfer among individuals, and in some cases establishing limitations on the powers inherent in ownership. Modern economic societies, whether socialist or capitalist, could not function without a well-developed system of property rights. The control and allocation of resources, which is part of the functioning of any modern technologically advanced society, requires such a system.[3]

For a variety of reasons, property rights may be imperfectly defined or found to be not fully enforceable. The person who owns a car can enforce his property rights in that car, for the most part. He can restrict its use to himself, he can exclude other people from receiving any benefits from the car, and he can protect the car from damage as he sees fit. If someone does run into his car, the law protects his property right by assigning liability and providing for the collection of damages. Similarly, if he runs into someone else, he is liable for the damages created.[4]

[2] For a more complete discussion of market failures and their effect on allocative efficiency—especially those of externalities caused by the absence of property rights and public goods—see Robert H. Haveman, *The Economics of the Public Sector,* New York: John Wiley and Sons, 1970.

[3] The major difference between Soviet-type socialist economies and capitalist-market economic systems lies not in whether property rights are defined but in whom the property rights reside, the state or individuals. Also, these systems differ in whether property rights are exchanged through markets or are controlled and allocated by some other institutional arrangements.

[4] The new no-fault insurance laws of some states can be interpreted as reducing the property rights of car owners by eliminating their rights to collect damages from the persons inflicting them. The economic justification for no-fault is that the costs of protecting and enforcing those property rights (lawyers fees, delays in court, court costs) were greater than the benefits.

However, with land, for example, other people may have their satisfactions increased or decreased depending on how the owner chooses to use his land. Land use has effects that spill over to other people besides the land owner. These effects are external to the economic decisions of the owner in that he has no economic incentive to take them into account.

Consider the following example. Mr. Brown owns a home on a piece of land adjacent to a plot owned by Mr. Smith. His satisfaction will be higher if his neighbor, Mr. Smith, uses his land for a well-designed residence rather than for a dump. Mr. Brown's higher satisfaction has a monetary counterpart, a willingness to pay. However, Mr. Smith's property rights do not include the power to extract payment from Mr. Brown for the pleasant view which a well-designed residence would create. Thus, Mr. Smith does not have an economic incentive to take Mr. Brown's wishes into account in deciding how to use his land. Alternatively, we could say that Mr. Brown's rights to his own property do not include the power to prevent Mr. Smith from imposing a loss of property value on him by creating a dump. And no mechanism exists through which he can make Mr. Smith compensate him for such a loss.

For another example, consider a river from which a brewery takes water to make its beer. If a paper mill locates upstream and dumps its wastes into the stream, it adversely affects the brewery. By its actions, the mill imposes costs or damages on the downstream brewery that it is not required to take into account. These costs are external to the mill's calculations. The mill has no incentive to economize on its use of the waste receptor services of the river.

Both the mill and the brewery feel that their location on the riverbank gives them rights to use the water in their different ways. In fact, these rights have not been clearly defined in law. Suppose the rights were originally vested in the brewery. Then the mill would have to bid for the right to use the river water. The mill would be required to buy the right to dispose of its wastes in the river by offering to pay more than the value the brewery derives from using the clean water for beer. The brewery's property right to the river, along with the legal systems developed to protect property rights, would prevent the mill from imposing costs on the brewery. On the other hand, if the mill owned the rights, it could sell clean water to the brewery and would do so if this produced a higher return on the asset than using it as a receptacle for its wastes.

The land and river examples have in common the fact that one party's actions affect a second party either favorably or unfavorably, but there is no requirement that the appropriate compensation payments be made. Without the payments, which are, in effect, prices—market signals, there are no incentives for the parties to take the appropriate or "best" action.

When externalities occur, they create a divergence between social values and private or market values. When a good or service is produced and sold in a market, the price reflects only those values that can be controlled by the buyer when the good or service is transferred to him. If third parties benefit from the existence of the good or the particular use to which it is put, its social value (including the value to third parties) is greater than its price or private value. Where such an external benefit is present, the third parties who benefit do not have to pay for the benefit that they receive. The result is a level of production of the good that is less than the optimum.

In Figure 4-2, the demand curve DD shows the marginal willingness to pay of the buyers. The intersection of this curve with the

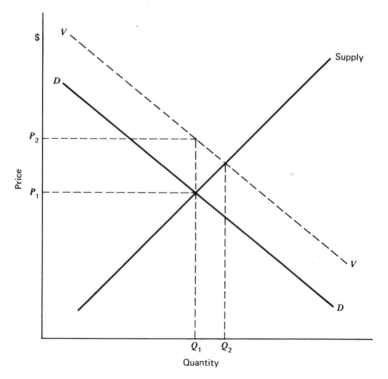

Figure 4-2 Market price and quantity with external benefits

supply curve determines marekt price (P_1) and quantity (Q_1). But if there are external benefits, the marginal social value curve (VV) is above DD. The difference is the amount the third parties would be willing to pay for those spillover benefits. At the market equilibrium, marginal social value exceeds marginal social cost, and the optimum output is Q_2.

If third parties are adversely affected by the production activity, they bear costs that are not included in the private cost calculation of the producer. Social cost (including third party costs) exceeds private cost. The result is excessive production and consumption of the good. In Figure 4-3 the supply curve SS includes only those costs of production borne by the producer. If the residuals from production impose external costs on downstream users of a river in the amount of E per unit of output, marginal social cost exceeds marginal social value (P_1) at the market equilibrium. In effect, consumers are being subsidized since the price they pay is less than social cost, and the subsidy is paid by the downstream users who bear the external cost. Production should be contracted to Q_2 where marginal social costs and marginal social value are equal.

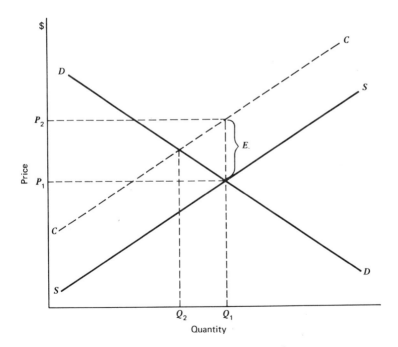

Figure 4-3 Market price and quantity with external costs

Third party effects, externalities, or divergencies between private and social values—terms that describe the same phenomenon—arise whenever property rights are not clearly defined or enforceable. This condition is almost universally applicable to environmental services. Consider the waste receptor services of the environment. If the discharge of residuals impairs other service flows (such as recreation) or causes damages (such as health impairment), there are third parties who are affected adversely. These parties are forced to bear the cost of the residuals discharge. But since their property rights in the environment are not enforceable, they cannot expect to be reimbursed. In the absence of any requirement to compensate those who are damaged by the dumping of wastes, the waste receptor service appears free to polluters. Since the price of the service is zero to them, they will expand its use to the point where the marginal value of the service is zero to them, even though the marginal social cost may greatly exceed zero. Hence, their use of the environment for residuals discharge will be excessive.

Public Goods and Market Failure

The other source of market failure is the public-good nature of many of the environmental services. A *public good,* once supplied to one individual, by its very nature is freely available to all. Since other users cannot be excluded because of nonpayment of the price, producers of public goods are unable to collect revenues from beneficiaries. Examples of public goods include aids to marine and air navigation—the ocean lighthouse is a classic public good—broadcast radio and TV signals, and national defense.

These examples represent the polar extreme of "pure" public goods. In other cases the "publicness" of a good is not an inherent characteristic, but arises because it is costly or inefficient to exclude those who have not paid the price. Examples are urban parks and streets. Many other goods have lesser elements of publicness about them. For example, there are public goods aspects to land use, since the owner cannot exclude his neighbors from enjoying the view of his beautifully landscaped property. In all cases where there is an element of publicness about the good, it will be provided in insufficient amount by private producers.

This line of reasoning is similar to the one provided above concerning divergence between social and private values. Indeed, public goods can be thought of as goods with an extreme case of external benefits. The social value of one unit of a public good, for example, one lighthouse, is the sum of the private values for all individuals. If the producer of the public good could obtain payment from all users, his revenues would be equal to the social value of the good. In this case, he would have an incentive to build more lighthouses as long as the marginal social value exceeded marginal social cost. But where publicness is a characteristic of the good, the producer is not able to capture the full marginal value, and hence, would lack the incentive to expand output to the optimal level.

Many environmental services have public good elements. Indeed, the benefits of clean air represent an almost polar case of a public good. Suppose a single firm that was polluting the air was granted exclusive property rights to the atmosphere. The firm could try to sell clear air, that is, reductions in pollution. But if it reduced pollution for one "customer," pollution would be reduced for all, whether or not they had paid. Since the firm could not capture the full willingness to pay for value of the clean air, it would have little or no incentive to curtail pollution.

Pollution and the Tragedy of the Commons

Professor Garrett Hardin has captured the essence of the environmental problem in the evocative phrase "The Tragedy of the Commons."[5] Hardin's example was a common grazing pasture into which each of the herdsmen in the village would introduce more cattle as long as there was a single blade of grass left. No one had any incentive to control entry and prevent overgrazing because the benefits of good management would not accrue to the person who withheld cattle. The tragedy was that all the herdsmen were locked into a system in which their individual self-interest directed each of them to take actions that were adverse, even disastrous, to all of them collectively.

The environment is the modern equivalent of the "commons." It is a *commonly held resource* by which we mean a valuable natural asset that cannot, or can only imperfectly, be reduced to private ownership. Examples are the air mantle, watercourses, complex ecological systems, large landscapes, and the electromagnetic spectrum. When the services of such commonly held resources are available at zero price and there are no other restrictions on entry or use, it is easy to predict the outcome. There is overuse, abuse, congestion, and quality degradation.[6]

---

[5] Garrett Hardin, "The Tragedy of the Commons", *Science, 162*, No. 3859 (December 13, 1968), 1243–1248.

[6] A freeway is an example of a resource that could be rationed by the use of prices (tolls), but that society has decided to treat as if it were a commonly held resource. During some periods, the result of free access is bumper-to-bumper traffic and severe congestion.

The materials balance principle has taught us that the mass of inputs to production must be equaled by the mass of outputs of residuals. While most of the economic activity associated with the flow of materials inputs and with production and "consumption" can be conducted through markets and exchange because private property rights exist, the residuals return flow goes largely to the commonly held environmental resource without passing through a market. These flows, therefore, lie outside the scope of the market system.

As far as we know, the laws of conservation of mass and energy have always held, but at lower levels of population size and economic activity, the return of "used" materials and energy to the environment had only localized effects. Most of these could be dealt with by means of ordinances and other local government measures to improve sanitation in and around the immediate vicinity of cities. As industrial deveolpment has proceeded, more and more material and energy has been returned to the environment, placing greater demands on its capacity as a waste receptor. The environment has become a scarce resource. Yet because private property rights in the environment have not been assigned, no market signals have been flashed to guide this unique resource to its best use. Larger "problem sheds" are being affected and greater numbers of people more remotely located in both space and time are suffering adverse impacts.

Commonly held environmental assets that cannot enter into market exchange are progressively being degraded because their use as "dumps" appears costless to the industries, municipalities, and individuals using them that way. This is so even though important values from other uses of the asset are degraded or destroyed. For example, the destruction of trout fisheries or increased mortality from respiratory disease nowhere enter into the profit and loss calculations of economic enterprises that get their signals and incentives from market prices. Since the "commons" are free, though valuable, no individual user has an incentive to husband them and protect them, and their increasingly degraded state yields a narrower and narrower range of services. For instance, at the limit, a watercourse may be capable of yielding no service except the carriage of wastes. When such things happen in an affluent market economy, we have witnessed a fundamental failure in the incentive system.

Not only do commonly held resources such as air and water get overused and misused, but also opportunities to improve their quality (increase their yield of services) are left unexploited. The ability of streams to receive waste waters without damage to other uses can be enhanced by raising low flows through reservoir storage and release and by mechanically introducing air into them. Degraded ecosystems can sometimes be improved—through introduction of exotic species, for example. Changing land use patterns and transportation systems can have favorable effects on environmental quality. But because of the public-good character of these activities, market exchange does not provide incentives for undertaking them. If they are to be done, it must be through collective investment and management. Also, in a more long-run or dynamic context, the absence of market exchange means that appropriate incentives are not provided for technological improvement with respect to the conservation and collective management of common property resources.

In summary, a great asymmetry has developed in the effectiveness and efficiency of our system of economic incentives. On the one hand, the system works well to stimulate the exploitation of basic resources and to process and distribute them, but it fails almost completely with respect to the disposal of residuals to commonly held resources. The result is that they are mismanaged, overused, degraded, and suffer from underinvestment and technological lag. An immediate corollary is that we have too much waste material and energy generated and disposed of to the environment, too little recycling, by-product recovery, and process design to save materials and energy, and, consequently, too rapid a rate of exploitation of basic natural resources.

We have identified the cause of the "tragedy of the commons" as widespread failure of the market system as an institution for allocating environmental resources. Hardin suggested that the solution was mutual coercion, mutually agreed on, in other words to replace markets with some governmental institution to make environmental management decisions and to implement them. This is not the only possibility. It is also possible to change the rules under which markets operate. Just as there may be inefficient markets, there may be inefficient governments. Identification of the cause of a problem is only the first step in the process of finding solutions. In Chapter 5 we consider alternative kinds of institutional changes that could be part of an effective pollution control strategy.

5

The Economic Principles of Pollution Control

We have seen how market failures due to lack of property rights and public goods lead to a misallocation of resources. We have also seen examples that suggest that market failure is common where environmental services are concerned. Understanding market failure and its attendant waste, we must now inquire how society can take collective action to correct the misallocation of environmental resources that is caused by this pervasive market failure. In this chapter we will first develop the concept of the optimum level of pollution control and contrast this with the level of pollution control attained in an unregulated market economy. Then several alternative pollution control strategies will be described and evaluated. The basis of the evaluation will be the extent to which the strategy will lead to the optimum level of pollution control at the lowest possible cost as well as other factors such as feasibility and overall effectiveness.

THE OPTIMUM LEVEL OF POLLUTION CONTROL

Pollution control absorbs scarce resources that could be used to produce other things of value. Just as we can be sure that today we have too much pollution, we can say that it is possible to have too much pollution control. This would be the situation if the resources used to obtain the last bit of pollution control could have produced other things of higher value.

How far we go in the direction of pollution control is a matter of choice, and rational choice requires some criterion for comparing and evaluating alternatives. Our criterion is *economic welfare*. But what do we mean by economic welfare?

In a society whose political and economic institutions are organized around the principle of the primacy of the individual, the aggregate welfare produced by the economic system derives somehow from the economic welfare of the individuals who comprise this system. In turn, the economic welfare of the individual depends on the quantities of goods and services (including environmental services) that he consumes. A principal problem in moving from the notion of an individual's economic welfare to the economic welfare of the society as a whole is the question of the distribution of welfare or equity.

If one person is made better off while no one else is made worse off, then there is a presumption that the society as a whole is better off.[1] But where one person's gain comes at another's expense, statements about better or worse must be based on some notion of the deservingness of the gainers vis-a-vis those who lose. The question is how we judge the way the economic pie gets sliced. This question cannot easily be separated from the question of the size of the pie. In comparing two alternatives, we might reject the one with the larger pie if it is distributed so that a very few wind up with large pieces while most get short rations.

We encountered the same problem in Chapter 4 in trying to evaluate the outcome of a competitive market system.[2] There we said that if it can be assumed that society had acted collectively through representative political processes to achieve what it considered to be a fair distribution of resources and income, then the outcome of a competitive market system could be considered to be optimal. This is because given the assumptions, competition yields the largest possible pie. Similarly, in what follows we assume that whenever the size of the economic pie is increased, no matter how it is distributed, economic welfare is increased. We do this not because we believe that the present distribution of income is just or fair, but only because this assumption provides a convenient way of attacking an important part of the problem of making choices about environmental quality. We will point out several instances where considering only the size of the pie to the exclusion of distributional questions could lead to bad public policy choices.[3]

If the criterion is the size of the economic pie, the next question is, "What goes into the pie?" The two components of the pie that are of interest are the output of goods and services for private consumption, government, and investment (N), and the stream of environmental services net of any environmental damages (E),

---

[1] But if economic welfare were already distributed very unequally and the person who was made better off already had a disproportionate share of the total, one could legitimately judge this to be a decrease in social welfare.

[2] See pp. 70–71.

[3] See especially Chapter 8, pp. 143–148.

both measured in dollars of constant value. Pollution has been defined as the impairment of the flow of environmental services by residuals discharges. Let E^* represent the value of the flow of services in the absence of any residuals discharge or pollution. Then the difference, $E^* - E$, is the damage due to pollution or D. Similarly, in the absence of any pollution control measures, the economy could produce N^* worth of goods and services. Pollution control would absorb scarce resources (T), reducing the flow of goods and services to N. This can be summarized as follows.

$$\text{Welfare} = W = N + E \qquad (1)$$
$$= (N^* - T) + (E^* - D)$$
$$= (N^* + E^*) - (T + D)$$

Only D and T are affected by the way in which residuals are disposed of. Their sum $(T + D)$ is the total cost of residuals disposal, and represents a reduction in economic welfare. The pollution control policy that maximizes economic welfare is the one that minimizes these costs.

From Equation 1, changes in economic welfare associated with pollution control are given by:

$$\Delta W = -\Delta D - \Delta T \qquad (2)$$

where $-\Delta D$ is a reduction in pollution damages. An increase in welfare requires a decrease in damages $(-\Delta D)$ holding treatment costs constant, a decrease in treatment costs $(-\Delta T)$ holding damages constant, or an increase in treatment costs that is more than offset by lower damages. Additional pollution control steps should be undertaken as long as the reduction in damages $(-\Delta D)$ exceeds the cost of achieving them (ΔT). For example, if an increase in pollution control reduces damages by $5 ($-\Delta D = \$5$) but costs only $3 ($\Delta T = \$3$), taking that step produces a net gain of $2 in economic welfare. If additional treatment expenditures bring diminishing returns, or equivalently, if additional damage reductions come at increasing treatment costs, then economic welfare is maximized by extending pollution control to the point where $-\Delta D$ equals ΔT.

This statement of the optimum condition assumes that dollar values can be attached to all of the environmental services and damages (losses in services) so that they can be added to national income to determine welfare. Since the concept of economic welfare is based on the welfare of the individual, the dollar values used must be based on the valuation of these services by the individual who receives them. The value of an increment to an environmental service is the maximum amount of money that person would be willing to pay to receive that increment rather than do without.[4] Where goods and services are exchanged in markets, the price is equal to the marginal willingness to pay.[5] But where there is market failure, prices are not accurate measures of the values of environmental services. In these cases techniques for inferring these values from other information must be developed. Some possibilities will be described in later chapters.

THE POLLUTION CONTROL MODEL

Models are abstractions from or simplifications of reality. These simplifications are not meant to distort our view of reality. Instead, by stripping away nonessential detail, the simplifications enable us to see the most important features of the problem at hand. The purpose of this model of pollution control is to make clear the relationships among pollution control costs, pollution damages, concentrations of residuals in the environment, and rates of discharge of residuals into the environment.

Finding the Optimum

First assume that there is only one harmful residual being discharged to the environment, and the damages caused by this residual depend only on the concentrations of this material in the environment. Further assume that we know the effects of various concentrations of this residual in the environment and can place dollar values on these effects or damages. These dollar values can be interpreted as the willingness to pay on the part of all people affected by this residual in order to *avoid* these damaging effects. In other words, these dollar values are the maximum aggregate willingness to pay to restore the environment to a natural or unpolluted state.

[4] This is the first instance where failure to consider the distribution of income can lead to bad choices. When resource allocation and environmental management decisions are based on willingness to pay, this is, in effect, a system of dollar voting. Those with the most dollars get the most votes. For example, if only the poor were damaged by air pollution, the willingness to pay (dollar votes) for pollution control would be very low.

[5] If price were less than marginal willingness to pay for an individual, he would increase his purchases of the good until diminishing marginal utility brought his marginal willingness to pay into equality with its price. For elaboration see Robert H. Haveman and Kenyon Knopf, *The Market System*, 2nd Ed., New York: John Wiley, 1970. See also Chapter 4, pp. 68–70.

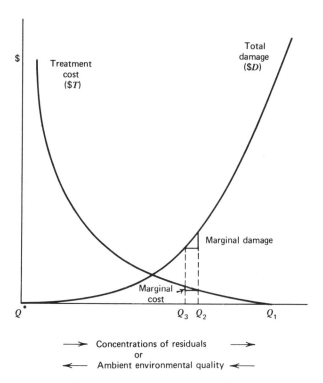

Figure 5-1   Damages and treatment costs for environmental quality

Figure 5-1 shows the relationship between willingness to pay or total damages ($D) and increasing concentrations of the residual in the environment. Moving to right along the horizontal axis represents increasing concentrations of the polluting material and increasing total damages. Conversely, moving to the left represents improving ambient environmental quality. Q^* represents the zero pollution state or maximum attainable environmental quality. In the absence of any pollution control efforts, the level of pollution (or environmental quality) depends on the level and nature of economic activity. Assume that at a certain point in time, the economic activity and its associated throughput of materials resulted in a level of pollution indicated by Q_1, in Figure 5-1.[6]

[6] The point Q_1 will tend to shift out to the right over time with growth in income and population.

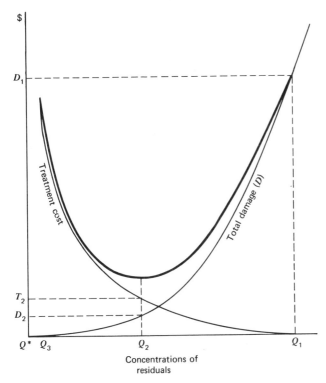

Figure 5-2   The optimum environmental quality

Pollution can be reduced and environmental quality improved by incurring pollution control costs. The curve sloping upward to the left from Q_1 represents the total cost of improving environmental quality. This curve is labeled "Treatment costs." In this case "treatment" is a shorthand term for all of the technological options described in Chapter 2.

According to the previous discussion, the optimum level of pollution control occurs where the total cost of residuals disposal is at a minimum. The total cost of residuals disposal is the sum of damages and treatment costs $(D+T)$. In Figure 5-2 this sum is shown by the heavy line. This line represents the vertical summation of the total damage and treatment cost curves of Figure 5-1 (also shown in Figure 5-2). This total cost curve has its minimum at point Q_2. Thus, Q_2 is the optimum level of environmental quality. This means that treatment costs equal to T_2 should be incurred in order to improve environmental quality from Q_1 to Q_2 and to reduce the damaging effects of pollution from D_1 to D_2. An expenditure of T_2 buys pollution control which is "worth" D_1 minus D_2.

The problem of finding the optimal degree of pollution control can be restated in terms of the marginal or incremental analysis as expressed in Equation 2. There is an incremental or marginal damage curve associated with the total damage curve that shows how much total damages increase with each small increment to the concentration of residuals. In geometric terms, the marginal damage curve is the slope of the total damage curve. This is shown in Figure 5-1. Similarly, there is a marginal cost of treatment curve associated with the total treatment cost curve. Both of these curves are shown in Figure 5-3. At points to the right of

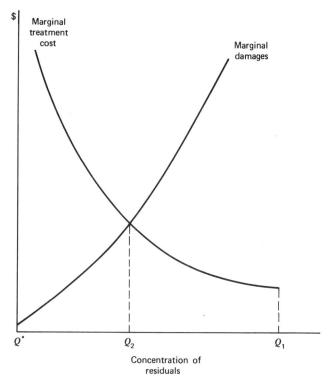

Figure 5-3 Marginal damages and marginal costs for environmental quality

Q_2, the marginal damage curve is above the marginal treatment cost curve, showing that a small reduction in residuals concentrations (a move to the left) will produce more in benefits (reduced damages) than it will cost to achieve it. Concentrations should be reduced as long as the marginal benefits (reduced damages) exceed the marginal costs of achieving them. The optimum degree of pollution control and the optimum concentration of residuals is at Q_2 where the marginal damage just equals the marginal treatment cost. Continuing to control pollution beyond that level will add more to the cost of treatment than it will deduct from the costs of pollution, showing that pollution control has been carried too far.

It must be recalled that this is optimal only in terms of the stated criterion, the maximum size of the economic pie. To insist on a higher level of treatment and lower residuals concentration would impose additional costs on those who bear the costs of pollution control. These costs would be larger than the benefits from the improved environmental quality. If the beneficiaries comprise a different group than those who bear the costs, they have every incentive to seek higher and higher levels of treatment right up to and including 100 percent treatment. Nothing less than 100 percent is optimal, at least in their eyes. But are the beneficiaries more deserving as a group than those who bear the costs? Actually our assertion that Q_2 is the optimal level of pollution control is based on the judgment that the dollar votes of the beneficiaries that are tallied by the total damage function should count equally with the dollar votes of those who bear the costs.

It is also true that those who must bear the cost of pollution control would prefer a zero level of treatment. Furthermore, lacking either incentives or compulsion to undertake treatment or residuals management costs, dischargers will minimize T by undertaking no treatment. Residuals concentrations would be at Q_1. Thus, in the absence of an effective pollution control strategy, the damages due to pollution or the value of environmental service foregone will be $\$D_2$. (See Figure 5-2.) As the curves show, a reduction in residuals concentrations from this point will add more to the realized value of environmental services ($E^* - D$ in Equation 1) than it costs in terms of reduced output; and the sum of output plus environmental services ($N + E$) will be increased.

The optimum in pollution control has been described in terms of the level of residuals in the environment. The criterion compared the value of reductions in the level of residuals with the costs of achieving them. It is also possible to restate the problem in terms of rates of discharge of the residual material. In order to do this we require information on the relationships between

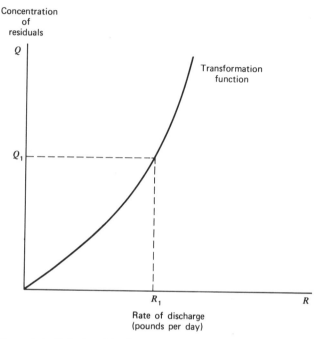

Figure 5-4   Relationship between residuals discharges and environmental
quality

rates of discharge of residuals and the resulting concentrations of
the materials in the environment. In practice these relationships
are quite complex, depending on a host of environmental parame-
ters, many of which vary randomly over time (for example, temperature, wind, stream flow) as well as the rate and time pattern
of discharges of the residuals.[7] For simplicity, however, let us
assume that the concentration of the residual material in the
environment depends only on the rate at which the material is
being discharged, and that this relationship can be portrayed by
a simple graph such as Figure 5-4. For each level of residuals discharge (for example, R_1), this *transformation function* shows the
ambient environmental quality or residuals concentration that
would result (for example, Q_1).

This diagram can now be combined with Figure 5-2 to show

---

[7] Recall the discussion in Chapter 3, pp. 48–50 and 54–56.

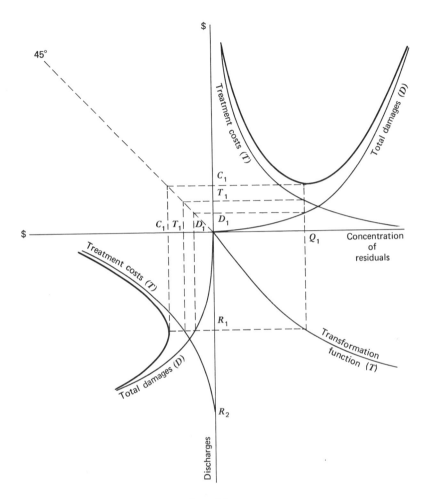

Figure 5-5 The pollution control model

the relationships among discharge rates, environmental quality, damages, and costs. See Figure 5-5, which takes advantage of the common axes of these two figures. The upper right-hand quadrant of Figure 5-5 is identical to Figure 5-2. The lower right-hand quadrant is Figure 5-4 rotated 90° clockwise. This relates discharges to concentrations of residuals. For example, if residuals discharges are R_1 the concentration of residuals in the environment will be Q_1. Now following the dotted line up from Q_1 to the damage function in the upper right-hand quadrant, then around counterclockwise to the lower left-hand quadrant, we find the dollar damages ($$D_1$) caused by the R_1 discharges. Similarly, we can find the cost of reducing residuals discharges to R_1. Every point on a curve in the upper right-hand quadrant corresponds to a similar point in the lower left-hand quadrant, and this point can be found in the same way by use of the transformation function and the 45° line. The optimum level of environmental quality, Q_1, corresponds to the optimum rate of residuals discharge, R_1. If the optimal environmental quality is known, the degree of discharge reduction necessary to achieve it can also be found.

Before describing how this model might actually be used by pollution control planners, we must digress to clear up two loose ends. The first concerns the treatment cost curve. This curve shows the lowest possible cost for attaining a given level of pollution control. It reflects the choice of the best or least-cost mix of technological options for control. If there is only a single discharger, this means that he has chosen his mix of technological options so as to equate the marginal costs of using all the options. If the marginal cost of reducing discharges through Option A were $10 while the marginal cost of Option B were only $1, the total cost of attaining that given level of pollution control could be reduced by $9 by expanding Option B. This shift of resources from one option to another should continue until there are no differences in the marginal costs of various options.

To achieve pollution control efficiently, each discharger must equate the marginal costs of all of his pollution control options, and each discharger's marginal cost must be equal to the marginal costs of all other dischargers. In other words, the minimum cost of achieving a given level of pollution control requires the equating of marginal treatment costs among all options and among all dischargers. An important aspect of pollution control planning is to try to satisfy this cost minimizing condition. If this condition is not met, the actual cost of any level of pollution control will be above that shown by the curve.

The second loose end concerns our assumption that there was only one polluting substance or residual. The existence of two or more residuals flows complicates the model in two ways. First, as the materials balance model showed, there are interdependencies among residuals in their generation and discharge. For example, a decrease in smoke or particulate emissions on the part of the firm may be accompanied by an increase in the discharge of solid matter into the nearby river. To take this into account the "treatment" cost curve for one residual (for example, smoke) must include the environmental damages caused by the increased discharge of the other residual (for example, solids). It is possible to construct models that take many interdependencies into account and solve them simultaneously for the optimum degree of control of all residuals flows.

The second way that the presence of two or more residuals can complicate matters is through interdependencies in the effects of residuals on the environment and its inhabitants. When there are synergistic relationships among pollutants, estimating and analyzing the damage curve is made more difficult. Information about the nature of these relationships is difficult to obtain experimentally, and the economic analysis is complicated by the fact that the damaging effect of one polluting substance depends not only on the level of that pollutant in the environment but also on the level of its synergistic partner.

Applying the Model

In order to make use of this pollution control model in undertaking the management of environmental quality, the decision maker must be able to obtain three kinds of information. First, he must determine the relationship between residuals discharge reduction and treatment cost (the treatment cost curve in the lower left-hand quadrant of Figure 5-5). Next, the planner must develop a model of the assimilation and dispersion capacity of the environmental system in order to determine the relationship between discharges and ambient environmental quality (the transformation curve of Figure 5-4). With these two pieces of information, the planner can determine treatment costs as a function of environmental quality. The third piece of required information is the relationship between concentrations of residuals and dollar damages due to pollution. With this, the upper right-hand quadrant of Figure 5-5 is completed and the optimum level of environmental quality can be determined. Alternatively, the transformation function can be used to determine damages as a function of discharges, and the optimum can be found in the lower left-hand quadrant. The results are identical and they require the same three pieces of information. This approach has the advantage of indicating directly the required reduction in residuals discharges. In Figure 5-5, discharges must be cut by R_2-R_1 to attain the optimum.

Some Limitations

The purpose of outlining this model has been to make clear the basic elements of the environmental quality management problem, not to make the problem appear to be more simple than it really is. In attempting to restore a note of realism to the discussion, we will first discuss *difficulties* in applying the approach embodied in this model, and then some of the conceptual *limitations* of the model itself.

Each of the three steps described here involves enormous difficulties in data gathering and analysis. Consider first the treatment cost function. In a river or airshed with a large number of dischargers, all of which have a wide range of technological options for controlling residuals, there are too many possible combinations for the pollution control planner to be able to estimate the *least cost* way of achieving a given level of pollution control to a high degree of accuracy. Also, the technology and cost factors are continuously changing due to innovation and economic growth. The relationships between discharges and concentrations in the environment are known for only a few forms of residuals, for example, BOD and dissolved oxygen in water, or sulfur dioxide in the atmosphere, and even these are crude models which in several cases have not been adequately tested for their accuracy. Finally, the presence of nonpoint sources makes both the estimation of control costs and the prediction of concentrations in the environment more difficult.

Estimating the damage function poses the most formidable difficulties of all since this involves the money values of services that do not pass through markets and do not have prices recorded for them. As we shall see in the next section, however, lack of knowledge of the damage function need not be a barrier to the application of this model.

One major conceptual limitation of the model stems from the randomness of nature and our lack of knowledge about the future. The natural phenomena that are represented in the transformation function have elements of randomness to them. For example, the relationship between BOD and DO depends in part on stream flow, which in turn depends on rainfall. Suppose we wish to build treatment facilities along a river with enough capacity to prevent fishkills from occurring even in the years of worst (lowest) rainfall and stream flow. If we base our plans and investments on the worst year observed in the past, we might be all right. Then again, next year may bring a record drought. If it does, we will have spent our money for nothing, since the luck of the draw will have killed our fish anyway.

Environmental management involves committing resources today to cope with expected future problems. It is inherent in the nature of things that the future may turn out to be different from our expectation. We run the risk of being surprised. The simple model presented here does not help us to see that risk, tell us who actually bears the risk, how the presence of risk affects decisions, or how we can buy insurance.

A second limitation has to do with possible cumulative and irreversible effects such as those discussed under the heading Global Problems in Chapter 3. The residual carbon dioxide being discharged to the atmosphere this year is not itself causing any environmental damages that we know of. There is the possibility, however, that it, in combination with the cumulative discharges of past and future years, could trigger major and essentially irreversible changes such as polar ice cap melting and flooding. Once those changes start, it is too late to control carbon dioxide emissions. For problems such as this, a model that focuses on present rates of discharge and present observable damages is not much help.

A third limitation stems from the sequential dynamic nature of the real world. The model assumes an unchanging static world in which the curves do not shift, and an optimal solution can be found and implemented. In the real world, the curves are in constant movement, and the process of change and sequence of policy steps may be more important than striving to attain the "optimum" at any point in time. For example, we presently have the choice of committing substantial resources to install the optimal air pollution control systems on automobiles; however, we may be better off fifteen years from now if instead, we devote those resources, to research and development of alternative transportation systems.

This discussion is not meant to imply that the model presented here is of no value or should be scrapped. Indeed we will continue to make use of it in discussing current air and water pollution policies in the next two chapters. Instead we want to emphasize that the model applies only to a subset of all possible environmental problems, and that other models and other analytical techniques need to be developed for looking at those problems that do not fit neatly into this framework.

THE ECONOMICS OF ENVIRONMENTAL QUALITY STANDARDS

Defining the optimum level of pollution control requires knowledge of both the pollution control costs and the damage costs. Control costs are the costs of resources used in carrying out the mix of technological control strategies. Since these resources are drawn from the market, their prices are recorded. As a result, control costs can be calculated and measured in dollar terms.

However, damage costs involve reduced flows of environmental services for which prices and values in dollars are not readily ascertainable. In the absence of knowledge of the real value of these damages, the optimum as defined previously cannot be identified.

As an alternative it is possible to approach the question of what level of pollution control we want in either technical or political terms. The technical approach is to utilize knowledge of the physical effects of residuals in the environment to establish maximum safe or acceptable pollution tolerances or minimum ambient *environmental quality standards*. For example, if it were known that concentrations in the atmosphere of sulfur dioxide above a certain level would cause detectable changes in the functioning of lungs of humans exposed to it for one hour, and this were judged unacceptable damage, then the ambient air quality standard would be set at that level. A standard such as this is equivalent to a vertical total damage function at that point. Pollution concentrations up to the level of the standard are implicitly assumed to have no cost; discharges resulting in concentrations above that level are assumed to have infinite costs. Although it is unreasonable to believe that the true damage functions have this shape, standards based on technical data can still provide a way of moving ahead in the absence of perfect knowledge about economic damages.

An environmental quality standard could also be established through some political choice process. For example, a tentative air quality standard could be "priced out" to determine the cost of achieving it. If the costs are judged to be too high (not very high), then the standard should be lowered (raised). If the standard is accepted, this implies that the community judges the damages avoided to be (at least) equal to the costs. In this way, any decision about environmental quality standards or goals, whether made on political or technical grounds, implies the monetary value placed on environmental quality. For example, one could interpret the single-minded pursuit of growth in GNP as implying the judgment that environmental services have zero value. Conversely, to take an extreme position, one who advocates that we all return to an eighteenth century pastoral way of life as a means of protecting the environment implicitly places a near zero value on material goods and services, or what amounts to the same thing, an extremely high value on environmental services. So, although some may argue that it is impossible to place dollar values on environmental quality, in reality we do just that whenever we make choices about how much pollution control we are going to try to attain.

The political and technical approaches for setting environmental quality standards and values cannot really be separated in practice. The choice of an "acceptable" level of damage could not be removed from political processes, nor could political choices be made without some information (perhaps conflicting) on the physical effects of different levels of pollutants in the environment.

What is the most appropriate method for making these environmental choices? By referendum? Legislation? or administrative discretion? This is a difficult question. If the choices are made through an open political process where all parties are adequately represented and special interest groups do not enjoy privilege or power, then one would be inclined to accept them as the basis for public policy. However, if these conditions are not met, the resulting environmental quality standards would reflect the disproportionate weight given to the interests of those with the greatest influence or political power.[8]

POLLUTION CONTROL STRATEGIES

The Bargaining Solution

For the moment let us further simplify the world by assuming that there is only one discharger and one person being damaged. One pollution control strategy is to grant enforceable property rights to one of the two parties and rely on market transactions to achieve the optimum. Given the assumptions of the pollution control model, exchange would result in the optimum level of pollution control irrespective of the assignment of property rights. For example, with property rights vested in the polluting firm, the individual inhaling the polluted air would have to offer to buy reductions in residuals discharges. He would make offers according to his marginal damage function, offering smaller payments for each additional unit reduction in discharges. The firm would be willing to supply these reductions, as long as the offer exceeded the cost of supplying them, that is, the firm's marginal treatment cost.

On the other hand, if property rights were vested in the individual, the firm would have to buy the right to discharge residuals into the environment. Starting at zero discharges, it would be willing to offer an amount equal to the marginal treatment cost

---

[8] This is analogous to the manner in which the unequal distribution of income can affect willingness to pay measures of pollution damages. See p. 83, n. 4. Some political dimensions of establishing pollution control policy are discussed further in Chapter 9, pp. 167–170.

avoided by permitting additional discharges. The individual would continue to sell rights to discharge as long as the offered payment exceeded his marginal damages for the increments sold. In both cases the "market equilibrium" would be the optimum where marginal treatment costs equaled marginal damages. The only difference between the two results would be in the resulting distribution of wealth or income.

There are two main barriers to greater reliance on the vesting of property rights and private exchange. The first is the size of the environmental units for which property rights must be vested. An air shed or a major river system are inherently indivisible. To secure the economic benefits stemming from the creation of property rights, control over the whole system must be granted to a single entity. Yet control over large environmental resources would convey enormous economic power to the owner. The result could be monopoly power and the associated misallocations of resources.

The second barrier is the presence of public good attributes in many types of environmental services. The inability to exclude those users who had not paid for the service would make it difficult if not impossible for the owner of the environmental resource to collect revenues equal to the willingness to pay for or the value of the environmental service being provided.

This is not to say that private bargaining and exchange will never take place with respect to environmental services. They may and it is worth discussing in more detail the conditions under which private exchange will work. The concept of *transactions costs* is essential for an understanding of the potential and limitations of private exchange with respect to the environment. Every transaction is costly in the sense that it takes time and perhaps other resources to locate and evaluate the opportunity for exchange, gather information, negotiate, and execute the transaction. If the transactions costs associated with private exchange are small relative to the potential gains from exchange, private exchange will take place.

In the case of the environment transactions, costs are likely to be quite high because of the many parties involved and the public good characteristics of environmental services that prevent or make difficult the exclusion of nonpayers. And where transactions costs are large, exchange will not take place even though there may be potential gains. (The presence of potential gains is signaled by the fact that marginal treatment costs are less than marginal pollution damages.)

The case for public intervention into the allocation of environmental resources is based on the presumption that the resource costs of governmental action (for example, gathering information, administration, and enforcement) are small relative to the private transactions costs that tend to impede a bargaining solution and small relative to the potential gains from exchange. We turn now to a discussion of several forms that public intervention might take.

Residuals Charges

First consider the case where the aggregate damage function is known. In this case, the state can attain the optimum level of pollution control even if it does not know the treatment cost function. The state could charge residuals dischargers for their use of the waste assimilative capacity of the environment. It would establish a charge or price equal to the marginal damage for each unit of residuals. The result would be equivalent to the two-party exchange described previously. The residuals charge would act as an incentive to economize on the use of the environment. Dischargers would decrease their residuals flows as long as the marginal cost of doing so was less than the charge for discharging, settling at the optimum where marginal treatment costs equal the charge (marginal damage). The sum of the treatment costs of all dischargers is minimized because each discharger is equating his marginal cost to the common charge—hence, marginal costs of all options among all dischargers are made equal.

It would be instructive to examine the impact of such a charge on the choice of technological options for controlling pollution. First, the discharger may be able to undertake treatment of part or all of his residual flow; but, he also may find that process changes, recycling, or materials recovery are less expensive than treating the wastes or paying the charge for discharging the residuals into the environment. Also, if the charges vary according to the time and place of discharges, for example, higher for urban areas or in critical low stream flow periods, the discharger will respond by choosing less damaging places for a given discharge or by halting or reducing his discharge during the periods of high charges. In any case, he will seek the least costly combination of treatment, recycling, process change, and discharge pattern.

Furthermore, since all of these options are costly, the price of his product will tend to rise to cover costs. In this way the costs of pollution and pollution control tend to be borne by those who purchase the good. The effect of the rise in price is to reduce the consumption of this product, and shift consumption toward other products not involving the same level of environmental damage or residuals charge. A charge also provides a positive incentive for research and development of new techniques of residuals management.

In summary, the charge has the effect of inducing dischargers and consumers to choose the least-cost combination of all of the technological options for pollution control, save one, augmentation of the environment through investment. This option normally requires direct action by the state since it may involve significant economies of scale or its benefits may have public good characteristics. Where opportunities for investing in assimilative capacity are present, efficient pollution control requires comprehensive regional pollution control planning. But residuals charges can play an important role in such plans. This point is discussed further in Chapter 6.

If the damage function is not known and the policy objective is to meet stated environmental quality standards, residuals charges can still be the basis of a pollution control strategy. If the aggregate treatment cost function is known, the appropriate charge can be determined directly. See Figure 5-6 in which the vertical standard S replaces the marginal damage curve of Figure 5-6. The environmental quality standard means that the concentration of residuals must be kept at or below Q_2. The correct charge is one that is equal to marginal treatment cost at the point where that curve intersects the environmental quality standard. Thus the charge is equal to T_1 per unit of environmental quality improvement. In order to impose this charge on the residuals discharger, it must be converted into a charge per unit of residual. The transformation function of Figures 5-4 and 5-5 can be used for this.

A system of residuals charges implies that property rights in the environment have been vested in the users of environmental services with the state acting as agent for the sale rights to the waste assimilative capacity of the environment. Alternatively, if property rights were held to be vested in dischargers, the state could act as agent for the people and "buy" reductions in dischargers. As before, the transactions would be based on the marginal damage function, but in this case, the state would pay an amount equal to the marginal damage avoided for each decrease in residuals discharge. In many respects the effect on dischargers would be the same as in the case of the charge. Since an additional unit of discharge would cause the discharger to forego receipt of the bribe, the foregone payment would be an opportunity cost associated with the discharge. The discharger would withhold the residual if the payment for doing so were greater than the cost.

The incentives created by the bribe system would be similar to

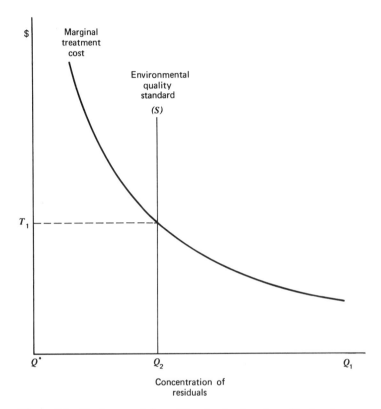

Figure 5-6 Environmental quality standard and residuals charges

those of the charge, and the dischargers' choices of technological strategies would be the same. However, there are some problems of administration that greatly diminish the attractiveness of the bribe as a practical alternative. For example, unless entry into the particular industry is strictly controlled, introduction of a bribe provides an incentive for new firms to enter the newly profitable industry of "production of residuals."

Regulation and Enforcement

The second management option open to the state is to regulate residuals discharges directly and use the police power of the state to enforce its regulations. This is sometimes referred to as the "regulation and enforcement" approach. The state could regulate the time and place of discharge and establish limits or *discharge standards* governing the allowable quantities and composition of discharges. The basis for establishing these discharge standards would be the known optimum level of environmental quality or the given ambient environmental quality standard. If the transformation function is known, the given optimum or standard can be translated into a total allowable quantity of residuals discharge.

Where there is more than one discharger, the state must then allocate this total among the dischargers on some basis; herein lies one of the problems with implementing discharge standards. Given the allowable total discharge, there is one allocation of this total among individual dischargers that will minimize the sum of all treatment costs incurred by all dischargers. Minimum total cost is achieved when the marginal treatment costs of all dischargers are equal at the allowed discharge levels. However, unless the state knows the treatment cost functions of all dischargers, its allocation of discharge permits will meet this condition only by accident. In the absence of cost information, the state would have to rely on some rule of thumb, allocating the discharge standards on the basis of production levels, or size, or perhaps equal standards for all.[9] Therefore, discharge licenses are likely to be allocated among firms in an inefficient and perhaps quite inequitable way resulting in higher than necessary total treatment costs.[10]

Since each discharger is free to choose its own technological strategy for meeting its standard, it will choose that set of technological options that minimizes its own cost of treatment. Also, as in the case of residuals charges, product price will increase because of the additional costs and consumption will be shifted toward goods with lower environmental costs. The increase in product price may be either more or less than would be the case under a residuals charge. The increase in price would be less in the event that the discharge permits were assigned so as to minimize aggregate treatment costs. This is because there is no resid-

[9] This point is illustrated with an example in Chapter 6. See pp. 112–115.

[10] If the state auctioned off the discharge permits to the highest bidders, the cost minimizing allocation would occur. But this is essentially a residual charge system. If the state issued the permits to firms, but then permitted exchanges of permits among firms, those with the highest marginal treatment costs would be able to purchase permits for additional discharges from other firms with lower treatment costs for whom the permits would have less value. This is another example of how prices can give signals for improving resource allocation. For further discussion of the concept of selling rights to pollute, see J. H. Dales, *Pollution, Property, and Prices;* Toronto: University of Toronto Press, 1968.

uals charge levied against the allowed discharge. However, if the allocation of discharge permits were inefficient, which is more likely, costs and prices would be higher—perhaps higher than would be the case with residuals charge.

Some Piecemeal Strategies

The two strategies discussed in the preceding paragraphs, residuals charges and direct regulation, are comprehensive strategies in the sense that the actions required of dischargers are, or at least can be, systematically related to the goal of attaining the optimum level of pollution control or the stated environmental quality standard. On the other hand, while piecemeal strategies may result in some decreases in pollution and improvements in environmental quality, they are at best only imperfectly related to the pollution control goals of the society.

One type of piecemeal strategy is the establishment of general regulations governing processes and other actions of dischargers. For example, some cities have placed upper limits on the sulfur content of fuel oils burned to generate electric power. The objective is to reduce sulfur oxides emissions. As long as fuel oil consumption is constant, total sulfur oxides emissions are directly related to the sulfur content of the oil. But if oil consumption increases, either because of the growth of demand for electric power or because increases in the prices of other fuels cause users to switch to oil, the growth in *total flow* of combustion gases to the atmosphere will at least partially offset the reduction in the *concentration* of sulfur oxides in the gases.

As another example, most states require that water wastes be provided with secondary treatment before discharge. Secondary treatment processes can vary in effectiveness under different circumstances. They can remove between 85 percent and 95 percent of the organic residuals from water. A slip from the upper to the lower end of the range means a threefold increase in the concentration of residuals in the effluent and, at a constant rate of effluent flow, a tripling of organic discharges.

Several specific criticisms can be leveled at this type of piecemeal approach. First, since the specification deals with processes rather than discharges, the effect of the regulations on the quality of the residuals flow, on total discharges, and ultimately on environment quality, is uncertain at best. Specifying percentages of residuals removal or transformation does not directly control the rate of discharge of residuals into the environment. And it is the latter that matters for environmental quality. If the percentage of a residual removed from an effluent is increased from 90 to 92 percent at the same time that the rate of discharge of the effluent increases 25 percent, the total quantity of residuals discharged per unit of time will be unchanged. Second, the state could never have enough information to determine for each discharger the least-cost technological mix for obtaining a given reduction in residuals discharge. Where similar regulations are imposed on several dischargers, differences in marginal treatment cost curves are likely to result in a higher than necessary total cost of achieving whatever pollution control is accomplished. And finally, once dischargers have complied with the regulations, they have no further incentive to take steps to manage residuals flows.

A somewhat different kind of piecemeal strategy involves offering positive financial inducements to discharges to undertake certain kinds of pollution control steps, for example, purchase and install abatement equipment. These inducements could take the form of either reductions in income or property taxes or cash payment to cover a portion of the cost. There are two serious disadvantages associated with the use of positive direct and indirect subsidies of pollution control activities.

First, for administrative reasons, the actions eligible for subsidy or tax benefits must be clearly defined. To prevent the policy from degenerating into a general subsidy of businesses, the eligible actions must be limited to specific treatment processes rather than the more general forms of process change, materials recovery, and consumption changes that might be part of the optimal technological strategy. Thus, firms are discouraged from investigating the widest possible range of alternatives and are more likely to choose a nonoptimal set of technological options to qualify for aid.

Second, since the subsidies and tax benefits do not cover the full cost of the actions, they must be coupled with some other form of sanction or incentive, (such as a residuals charge) to have a significant impact. In fact, it would be preferable to view this set of strategies as a way of sharing the economic burdens associated with pollution control measures brought about by some other pollution control strategy rather than as a separate strategy for attaining pollution control.

THE ROLE OF ENFORCEMENT IN POLLUTION CONTROL

All of the pollution control strategies require enforcement to some degree. Since they all require some action on the part of dischargers—reporting and paying of the charge, limiting of discharges, or installing of equipment—dischargers must be investigated, violations detected, and the judicial system used to impose penalities. Any pollution control policy must be backed up by an enforcement system. However, voluntary compliance reduces the need to invoke sanctions. The extent of and the demands placed on the enforcement system depend on the cost of compliance, the penalities associated with noncompliance, and the probability of a violation being detected. For example, if all polluters were drawn and quartered the day that their violations were discovered, violations (or at least discovery of violations) would be rare, indeed. Or, if pollution control were costless, the cost of avoiding penalties would be zero, and penalties would never have to be invoked.

There are four reasons for attempting to minimize reliance on the enforcement system and encourage voluntary compliance with whatever policies have been established. First, effective enforcement requires information, investigation, and resources devoted to the detection of violations and their prosecution in court. All of these are costly, and these costs are as much a part of the total cost of pollution control as constructing treatment plants. These enforcement costs are analogous to the transactions costs associated with private exchange. Other things equal, we should adopt strategies with lower enforcement costs.

Second, activating a legal enforcement system can be a cumbersome way of dealing with pollution problems. Making noncompliance with a pollution regulation a crime means that persons charged with that crime must be granted certain legal rights and procedural safeguards. The impact is to make enforcement a slow, drawn out and uncertain procedure. Neither the legal offices of our enforcement agencies nor the courts are equipped to handle the kind of case load that would result from bringing charges against all residuals dischargers. Nor are we endowed with the patience that would be required to see the whole process through.

There are apt analogies in our systems of personal income taxation and military conscriptions. We depend on voluntary reporting of income and payment of taxes. If there were widespread failure to report and/or refusal to pay, the system would break down. The United States and Great Britain have been successful in inducing voluntarism as the dominant mode of behavior with respect to income taxation, but Italy and France have not. In those countries, tax officials have essentially given up on the task of collecting taxes from their upper income groups. Lack of voluntary compliance with pollution control regulations and the inability of enforcement agencies to cope with widespread resistance to control efforts have led to a similar breakdown in our system of environmental control.

Voluntary submission to the draft was almost universal before

1967. Since then, however, increasing numbers have been going to Canada or fighting induction orders and classifications through the Selective Service appeal system and the courts. The increasing administrative and judicial burden imposed by this loss of voluntarism has been a significant factor in the move toward a "volunteer army."

The third reason for attempting to minimize the role of enforcement concerns the potential rewards for noncompliance. To the extent that enforcement actions play a significant role in pollution control policy, they may create perverse incentives for dischargers. Appealing a finding of a regulatory agency or seeking a stay of a court injunction may be cheaper for a discharger, even in the long run, than taking steps to control residuals. Hiring lawyers and lobbyists may be a better investment than pollution control equipment.

The final factor to be considered is the political nature of the whole process of setting regulations and enforcing them. The basic point is that compliance can be obtained either by compulsion (at gunpoint, as it were) or by "fair exchange."[11] As with relations between countries, clear superiority of force can enable the agency to play tough. But with limited investigative resources, procedural and legal safeguards, and an overcrowded court system, clear superiority of force cannot be established where heavy reliance is placed on the enforcement system. The alternative is negotiation and bargaining where the outcome depends on the relative strengths and talents of the two sides, but where both sides must give something in the process.

It can be argued that with such things as better drafted laws and regulations and more money for the agency, this system can be made to work reasonably well. But we believe that the enforcement process is inherently weak and not capable of handling the burden that would be imposed on it by the strategy of direct regulation.

The Direct Regulation Strategy and Enforcement

It is in the American tradition to create regulatory agencies to deal with problems caused by malfunctions in the economic system. The existence of agencies such as the Interstate Commerce Commission, Federal Power Commission, and Federal Drug Administration is sufficient to convince most people that the problems for which these agencies were created are being dealt with suc-

[11] See Matthew Holden, Jr., *Pollution Control as a Bargaining Process,* Cornell University Water Resources Center Publication No. 9, Ithaca, 1966.

cessfully, but a careful analysis of the evidence shows that this is rarely the case.[12]

The naive view of the regulatory process is that the agency establishes rules and regulations to govern the behavior of the regulated and to further the public interest. The threat of sanctions is thought to be sufficient to deter violations; but if any occur, violators, it is believed, are quickly brought to justice. The reality is quite different. Regulatory agencies have substantial discretion concerning the interpretation and application of their rule-making and enforcement powers, but limited power to compel or coerce the parties they regulate.

As a consequence, regulation/enforcement becomes essentially a political process entailing bargaining between parties of unequal power. In this process, the real issues are camouflaged in technical jargon, and the regulators are largely protected from political accountability for their actions. The regulatory agency and the interests they regulate bargain over the regulations to be set. They bargain over whether violations have occurred, and if so, who was responsible. They bargain over what steps shall be taken to correct infractions. In the rare instances where the bargaining process breaks down and the conflict moves to the courts for resolution, the judicial system seeks acceptable solution or reasonable compromise. Only rarely is it forced by the flow of events into making either/or choices. At every stage of this multilevel bargaining process, those being regulated have a lot at stake, while the public interest is diffuse, poorly organized, and poorly represented. Predictably the bargains struck favor those being regulated.

The regulatory/enforcement approach to pollution control is well characterized by this description. This strategy pits the power of pollution control authorities against the power of the polluter in an unending sequence of skirmishes and battles over licensing and the enforcement of regulations. As a consequence, the enforcement process is long and drawn out and often inconclusive.[13]

Residuals Charges and Enforcement

One of the more attractive aspects of a residuals charge strategy is the smaller burden it would place on its enforcement system. Given that environmental quality standards have been set, a per-

---

[12] For example, see Paul MacAvoy, ed., *The Crisis of the Regulatory Commissions*, New York: W. W. Norton, 1970.

[13] For examples, see John C. Esposito, *Vanishing Air, The Ralph Nader Task Force Report on Air Pollution*, New York: Grossman, 1970, and David Zwick and Marcy Benstock, eds., *Water Wasteland, The Ralph Nader Study Group Report on Water Pollution*, New York: Grossman, 1971.

unit charge would be levied on each of the prominent harmful substances found in effluents or emissions. Each discharger would be responsible for monitoring his discharge, reporting its composition and quantity to the public authority, and making the appropriate payments. There is an obvious comparison with the system of reporting and paying corporate income taxes. As in the case of the income tax, rules and standards of accuracy for measurement would have to be specified and audits for compliance would have to be undertaken. Dischargers would be required to install and maintain the required monitoring equipment, subject to audit and calibration for accuracy by the authorities. Information on discharges and payments would be recorded and made available for public scrutiny.

Such a system would reduce the scope for administrative discretion and bargaining. The primary decision is the level of the charge to be imposed, and this decision is a significant and highly visible one. There is also a performance criterion by which to judge the correctness of the decision on the rate of charge. If environmental quality standards are being met, the rate is high enough; if not, the rate should be raised. Finally, while the history of regulation suggests that the zeal and effectiveness of the regulatory agency diminish over time, the effect of a residuals charge is durable. It remains effective unless the real cost of a fixed charge is eroded by inflation or unless there is an explicit political decision to remove it.

CONCLUSIONS

In Chapter 6 and 7 we discuss and evaluate present and proposed alternative water and air pollution control policies. After assessing the effectiveness of present policies and considering some of the problems of implementing effective pollution control programs, we will make some suggestions for changes in policy based on greater reliance on residuals charges. Even at this point there are some tentative conclusions that emerge from our discussion of alternative pollution control strategies.

First, because of the economic incentives it creates and the decentralized decision making with respect to how much each discharger reduces his discharge and how he does it, a residuals charges strategy will yield a given level of pollution control at lower total economic cost than alternative strategies. Second, policy makers using a residuals charge strategy need less detailed information on individual dischargers to achieve their pollution control goals at the lowest possible cost than if they were relying on some alternative strategy such as direct regulation. Third, because collecting the charge is an administrative matter similar to taxes, a residuals charge strategy reduces the burden on the judicial enforcement system—thus reducing the costs of enforcement. Finally, the residuals charge creates a direct, powerful, and continuing incentive to control discharges as compared to the weak and sometimes perverse incentives inherent in other strategies.

6

Public Policy for Water Pollution Control

Chapter 5 described the application of economic analysis to the question of environmental quality. In that discussion, several commonly held notions about the environment and its quality were seen to be at odds with economic logic. Calls for the elimination of pollution and for pristine clear rivers and streams, for example, were seen to be at variance with the economist's notion of the optimum degree of pollution and the optimum amount of pollution control. Indeed, in that discussion we saw that it is possible to have too much pollution control! That discussion was based on the proposition that the environment is not unlike other economic resources such as land and capital. It presumed that the environment exists to serve the needs of man. Hence, like these other resources, the environment must be managed to this end. Management of the environment is, however, a complex process that requires much information and a sensitivity to the subtle and long-range effects of human action.

Management entails setting goals or targets, and planning and carrying out steps to achieve those goals. In the case of the environment, goals could be established after a comparison of the marginal benefits and the marginal costs of pollution control. Since it is difficult to quantify and attach dollar price tags to benefits, it is usually more practical to establish environmental quality goals or *standards* by political means. Once such standards have been set, the economic problem becomes one of achieving these objectives at minimum cost. While we consider the important matter of benefit measurement briefly below, most of the discussion in this chapter is concerned with policies to achieve water quality standards, and the costs of alternative policies. Although two alternative policies might succeed in reducing the pollution of a river, one may be substantially more costly than the other. These costs must be considered in choosing from among the available environmental management policies.

Although this chapter and Chapter 7 consider the air and water pollution control problems separately, we must remind ourselves that there may be important interdependencies between them.[1] In Chapter 2 we explained such interdependencies in terms of the materials balance concept. Here we may remind ourselves that, for example, when waste water is treated there is no destruction of materials. A semisolid sludge is removed that may be incinerated thus possibly contributing to an air pollution problem. In fact, since the treatment process itself requires material inputs and energy, the total mass of materials flow is increased by treatment. The objective of treatment is to transform materials into forms that can be deposited in the environment with less damage or that may be more readily confined (as in a landfill).

THE BENEFITS AND COSTS OF IMPROVING WATER QUALITY

Consistent with the lessons from our discussion of economic theory, we must first inquire into the benefits and costs of improving water quality. The benefits from cleaner rivers, lakes, and streams are relatively easy to identify but quite difficult to measure and evaluate quantitatively. The major identifiable benefit from improved water quality is the greater availability of water-based recreational services yielded by the resource. With cleaner rivers and lakes, boating and fishing could be expanded, and swimming would once again be possible, and an attractive idea. Those who would take advantage of these services would be better off. In fact, when made aware of the expanded opportunities for recreation, the beneficiaries could, in principle, be made to pay a sum of money for these opportunities rather than do without. Their willingness to pay is a measure of the benefits of the cleaner river.

In addition, there are two other forms of possible pollution control benefits. Improved water quality may yield benefits due to better commercial fishing. As pollution control efforts lead to a reduction in the residuals discharged to rivers and lakes, fish stocks grow and the costs of "harvesting" them fall. Also, there may be benefits for people who live in cities and towns that obtain their drinking water from the river. With a cleaner river, the costs to cities for treating and purifying the water will be reduced. These reductions in costs also represent social benefits. What evidence there is suggests that these two types of benefit are small in comparison with recreation benefits.

Although economists and others have not been notably suc-

---

[1] And indeed with solid wastes.

cessful in measuring the value of these benefits for improvement in the quality of particular rivers, progress is being made. One interesting approach to this measurement problem was presented in a study that focused on the Delaware River from Trenton to the sea.[2] In beginning their study, the authors state: "There is sufficient evidence that upstream users have reduced the quality of the water in the Delaware River from Trenton to the sea to such a degree that it has been made almost unusable for water recreational uses." The authors then focus on 1990 and seek to estimate the *increase* in the number of days of usage of the river for swimming and boating that would occur in that year if the quality of the river is improved so as to encourage these activities. After projecting increases in the regional population to 1990 and alterations in its income and socio-economic characteristics, the study concludes that in that year a cleaner river would lead to over 600,000 more occasions of boating per year and 60,000 more man days of fishing per year than would otherwise occur. If these boating and fishing opportunities are valued at, say, $3 each, the benefits from improving the quality of the river will be about $2 million per year by 1990. This estimate does not include the benefits for swimming, a major potential recreational use of the river. After comparing these values with the costs of improving the quality of the Delaware Estuary, the authors suggest that these benefits alone justify the improvement in water quality required to encourage these activities.

In comparison with benefits, the costs of pollution control are, at least in principle, easier to estimate. These are the labor, capital, and materials that go into building and operating the treatment plants, recycling systems, and so forth. The major problems in estimating the future costs of pollution control are predicting which technological options will be chosen by dischargers and knowing how much treatment or control will actually be undertaken. The President's Council on Environmental Quality has estimated the total costs of achieving the water quality standards which have already been established for our rivers. These estimates are shown in Table 6-1.

Of the total $38.0 billion, just over half ($19.2 billion) represents capital investments in facilities that will continue to serve the nation in the years after 1975. The remainder represents annual operating and maintenance costs of about $3.1 billion that will have to be met every year.

[2] Paul Davidson, F. Gerard Adams, and Joseph Seneca, "The Social Value of Water Recreational Facilities Resulting from an Improvement in Water Quality: The Delaware Estuary," in Allen V. Kneese and Stephen C. Smith, eds., *Water Research,* Baltimore: Johns Hopkins Press, 1966, pp. 175-224.

Table 6-1 The Costs of Water Pollution Control, 1970–1975
(in Billions of Dollars)

| | Total requirements over the 5 year period 1970–1975 | | |
	Capital Investments	Operating Costs	Total
Public			
Federal facilities	0.3	1.3	1.6
State and local treatment plants	13.6	9.3	22.9
Total	13.9	10.6	24.5
Private			
Manufacturing	4.8	7.2	12.0
Other	0.5	1.0	1.5
Total	5.3	8.2	13.5
Total	19.2	18.8	38.0

Source. The President's Council on Environmental Quality, *Environmental Quality—1971*, Washington, D.C., 1971, p. 111.

The data in Table 6-1 reveal two things of interest. First, although domestic wastes account for only between one-fourth and one-sixth of the present total discharges of organic materials, the estimated cost of waste treatment by state and local governments is about twice that for private manufacturing. There are three factors that probably help to explain this anomaly. First, unit costs of control for manufacturing plants may be lower because of the wide variety of technological options to them, for example, recycling and process changes. Second, federal policy has required that all public treatment facilities provide secondary treatment, whether or not it is required to meet water quality standards, while less strict control requirements have been imposed on manufacturers in some instances where lower treatment levels are consistent with achieving water quality standards. Third, a substantial number of manufacturing units are already or will soon be connected to public sewer systems. So, a significant but unknown portion of the $22.9 billion expected to be spent by state and local governments will actually be used to treat the wastes of industry.

The second point of interest is that for the public facilities, capital costs are higher than operating costs over the five year period, while the opposite is true for manufacturing. This can be explained by the public policy decision to rely on conventional secondary treatment processes for municipal wastes. Industry is relying to a much larger extent on internal plant changes that do not require large capital investments, but that raise plant operating costs.

ACHIEVING STANDARDS AT LEAST COST

Let us consider as a hypothetical example a river that has eight firms located along its banks. Each firm deposits its wastes into the river, with the quantity of waste discharged per day shown in Table 6-2. The total waste deposited in the river is 2400 pounds

Table 6-2 Waste Discharges and Marginal Costs of Reducing Waste Generation for Eight Firms Located on a Hypothetical River

Firm	Pounds of Waste Discharged per Day	Marginal Costs of Reducing Wastes per Pound, in cents	Wastes Discharged with 4.5¢ per pound Effluent Charge
A	100	2	0
B	200	5	200
C	500	10	500
D	400	4	0
E	400	8	400
F	100	4	0
G	200	2	0
H	500	10	500
Total	2400		1600

per day. Further assume that all firms' wastes have the same effect on water quality. The table also shows the marginal costs (assumed to be constant) that the firms will have to incur to reduce waste discharge.

Assume that a political decision has been made to raise the water quality to a certain level or standard. The responsibility for regulating discharges by firms to achieve that standard has been assigned to a pollution control authority. In order to attain the stream standard, the waste load must be cut from 2400 pounds per day to 1600 pounds per day. The question to be answered by the authority is: How is the required decrease in waste loads to be allocated among firms? One possibility is for the authority to impose and enforce discharge regulations, in which each firm is forced to reduce its waste loads by 100 pounds per day. Call this alternative "uniform reduction." What is the total cost to society of achieving the stream quality standard by this method? It is the sum of the costs to each firm (2¢ × 100 + 5¢ × 100 + . . .), or a total of $45.00.

Another way would be for the authority to force each firm, again through discharge regulation, to reduce its waste loads by one-third. This is a "uniform treatment" or uniform percentage removal strategy. It will also achieve the stream quality standard (assuming firms comply), but at a total cost of $56.00 [⅓ × (100) × (2¢) + ⅓ × (200) × (5¢) + . . .].

Finally, it would be possible for the authority to set a charge on the use of the stream for waste disposal. This effluent charge would have to be paid on every pound of waste deposited in the river by a polluter. The charge to be set must be high enough to achieve a waste load reduction of 800 pounds per day. By inspecting the table, it can be seen that any charge greater than 4¢ per pound would achieve the desired reduction. If, for example, a 4.5¢ charge were levied, firms A, D, F, and G would find it cheaper to reduce discharges than to pay the effluent charge, while firms B, C, E, and H would find it cheaper to pay the charge. The desired 800 pound reduction of waste discharges will be attained at a total treatment cost of $26—the minimum cost possible to achieve the desired waste load reduction.[3] The comparison of costs is:

Uniform reduction	$45
Uniform treatment	$56
Effluent charge	$26

At an effluent charge of 4.5¢, the authority would also be collecting $72 in revenues that could be used for other means of improving the quality of the stream. The firms paying these charges are in effect paying a rent for their use of the environment. Note, however, that although this rent is a financial cost to firms, it is not a cost to society, only a transfer payment.

The lesson from this example is a very simple one. The use of effluent charges attains any specified waste load reduction at

[3] In the contrived example shown here, the effluent charge caused some firms to reduce waste discharges completely and had no effect on the waste loads of other firms. This result occurred because we assumed a constant cost per pound of reducing discharges. In reality, each firm has a schedule of costs for waste discharge reduction. The cost of the first few units of waste load reduction is very low but as reductions of 60, 70, and 80 percent occur, the costs of reducing waste loads by one more unit becomes very high. With this kind of cost schedule, an effluent charge would cause all firms to reduce wastes by some amount with the low cost firms reducing discharges more than the firms which face high waste reduction costs.

minimum real cost while at the same time giving all firms a continuous incentive to seek lower cost means of reducing waste loads to avoid paying the charge. Could a policy of directly regulating discharges have achieved the same stream quality standard by selectively imposing discharge reductions in the same pattern as resulted from the effluent charge? Yes. However, the regulator would have to know the waste discharge cost of *all* of the firms in order to select the least cost pattern. In fact, these costs are unknown (and unknowable) to public agencies. The advantage of the effluent charge is that it *automatically* secures the least cost pattern of waste load reductions for any total reduction required. It does so by allowing each waste discharger to make a free decision on how to minimize his own costs, given that he is confronted by the charge on dumping wastes into the river.

This example has assumed that the only available technological options involved treatment at the source. But if other options are possible, such as treatment at regional facilities to take advantage of economies of scale, then the least cost solution would be different. Suppose, for example, that firms D, E, and F were close enough together so that it was economical to pipe their wastes to a centrally located treatment plant. If economies of scale were such that these wastes could be treated at a cost of, say 2½¢ per pound, the water quality standard could be met, even slightly exceeded, at an even lower total cost, since no other firms would have to treat any wastes. An important function of a pollution control authority is to identify cost saving possibilities such as this and to act accordingly.

Several studies of the Delaware Estuary have shown that efforts to identify the least costly means of attaining water quality standards can pay off handsomely. All of these studies were based on a computer simulation model of the river that predicted the pattern of dissolved oxygen along the river for given discharges of organic wastes. One study used estimates of the costs of waste treatment at each source to calculate the cost of achieving different water quality standards by alternative policies.[4] The policies studied were (1) the uniform treatment of wastes by all discharges, (2) a single effluent charge in the basin, and (3) a zone effluent charge in which the charge imposed is different in the various portions of the river depending, roughly speaking, on the assimilative capacity of the river in that zone. Some of the results are presented in Table 6-3.

[4] Edwin L. Johnson, "A Study in the Economics of Water Quality Management," *Water Resources Research, 3* No. 2, (2nd quarter, 1967), p. 297.

Table 6-3   Costs of Improving the Water Quality of the
Delaware Estuary under Alternative Policies
(Millions of Dollars per Year)

Water Quality Level	Uniform Treatment	Single Effluent Charge	Zone Effluent Charge
Modest Improvement (2 ppm of dissolved oxygen)	5.0	2.4	2.4
Major improvement (3-4 ppm of dissolved oxygen)	20.0	12.0	8.6

Source. Edwin L. Johnson, "A Study in the Economics of Water Quality Management," *Water Resources Research, 3* (2) (2nd quarter, 1967), p. 297.

Two things are apparent. Effluent charges can lead to substantial economies—50 percent or more, in comparison with the uniform treatment alternative. Under any alternative, the marginal cost of higher water quality can be substantial.

A more recent study based on the same model has examined the cost savings possible from constructing large-scale regional treatment plants.[5] The results showed that if three large, publicly owned plants were built, and they charged the users the full marginal cost of treatment, the total cost of pollution control in the Delaware region could be reduced another 40 percent. Not all the dischargers along the river would be connected to these plants, and those that were not would still be faced with the economic incentive of the effluent charge. Those that were connected to a regional facility would also find that the user charge they had to pay would give them an incentive to use process change or use pretreatment to reduce the quantities of wastes they sent to the public facility; they would do so as long as the marginal cost were less than the user charge they had to pay.

THE PRESENT FEDERAL POLICY

Present federal water pollution control policy has two main elements. The first is a program of federal subsidies to cities for the construction of waste treatment plants. The second is a procedure for establishing regulations limiting discharges and for enforcing

[5] Federal Water Quality Administration, *Mathematical Programming for Regional Water Quality Management*, Washington, D.C., 1970.

these rules through the police power of the state and ultimately through the courts.

Subsidies

The Water Pollution Control Act of 1956 established the first federal subsidy for treatment plant construction. This subsidy takes the form of federal grants to municipalities that cover up to 55 percent of the cost of plant construction. In addition, some states augment the grants to the point where cities are responsible for only 15 percent of total construction costs. Since 1956 this program has grown rapidly until today nearly $1 billion per year is being spent by the federal government to subsidize public waste treatment. To be eligible for these grants, a state must have adopted a plan for achieving water quality standards which is acceptable to the Environmental Protection Agency (EPA). Present regulations stipulate that such plans must require a minimum of secondary treatment (85 percent removal of organic wastes) or its equivalent.

This program encourages cities to provide at least secondary treatment of the wastes of all dischargers connected to municipal sewer systems. While these discharges include the bulk of the nation's households and commercial enterprises, the majority of the nation's industrial wastes are discharged directly to rivers and streams. To encourage waste treatment activities by these dischargers, the Tax Reform Act of 1969 allows accelerated depreciation of waste treatment plant investments for tax purposes. The objective of this $120 million annual "tax expenditure" is to stimulate more spending on pollution control by reducing the after-tax cost of such investments.

Regulation/Enforcement

In addition to these subsidy programs, federal policy provides for regulations to govern the disposal of wastes into rivers and mechanisms to enforce these rules. The first provision for federal enforcement actions against water polluters was contained in the 1956 Water Pollution Control Act. The law relied primarily on voluntary compliance stemming from abatement conferences attended by polluters and government officials. It usually proved impossible to get beyond the conference stage, and only one case ever got to court.

The Water Quality Act of 1965 assigned primary responsibility for implementing pollution control plans to the states. It requires states to establish water quality standards for their waters and to develop a program for attaining them. This program is to be regarded as a benchmark for judging the progress of a state in attaining its water quality standards and for assisting federal officials in determining when and where to undertake federal enforcement actions. In implementing this program, state agencies must first determine the maximum amount of discharges consistent with the water quality standards. Then they issue licenses limiting discharges in aggregate to this maximum, usually following some rule of thumb such as uniform percentage removal. In enforcing these license provisions, states must undertake surveillance of dischargers and must initiate judicial or quasijudicial proceedings when violations occur.

The 1965 law also authorizes federal enforcement actions whenever it is found that state-established water quality standards are being violated. Alternatively, a governor or state agency can request the EPA to initiate enforcement efforts to deal with an interstate pollution problem. The EPA can initiate court actions 180 days after notifying violators. This provision of federal law was not used at all until August 1969, and as of the end of 1971, the EPA had issued only twenty-seven notices.

The decision to proceed to litigation ultimately rests with the administrator of the EPA. Because of the lack of guidelines for deciding when a pollution problem is sufficiently serious to warrant legal action, this decision becomes a political matter. Thus, enforcement at the federal level is in practice far from comprehensive or uniform. While the large municipality or firm may confront a probability of ultimate legal action, the small polluter is virtually immune from such action.

Federal and state pollution control efforts have been limited in scope and coverage in two respects. First, they have been primarily aimed at organic pollutants and have given relatively little attention to nonorganic pollutants such as plant nutrients, heavy metals (for example, mercury), toxic materials, and heat. This is not so much because of any limitations in either federal or state law as it is because organic pollution is the most noticeable in its effects and relatively easy to deal with, given present technology. Now, however, increasing attention is being given to the more exotic forms of pollution.

Second, pollution control efforts to date have been limited to discharges from point sources, for example, factories and sewer pipes, and have ignored pollution arising from erosion and siltation, agricultural fertilizer, irrigation water and pesticide runoffs, and other nonpoint sources. This is an area in which neither technology nor policy is well developed. Yet, as point sources are brought under better control, nonpoint source pollution will loom as a larger problem, both relatively and absolutely.

The Assessment—A Failure

The assessment of both these aspects of existing policy must be a gloomy one. If the objective of policy is the improvement of the quality of the nation's rivers, as it surely must be, the existing strategy is a dismal failure. In 1969, the General Accounting Office released the results of a detailed study of several rivers. That report concluded that even though $5.4 billion had been spent at all levels of government for waste treatment plant construction during the previous twelve years, the nation's rivers were in worse shape than ever before.[6]

This record of failure is attributable to both the subsidy and the enforcement aspects of current policy. Consider the waste treatment grant program. Because of the structure and administration of that program, much was spent but little achieved. The following are some of the reasons for this result.

1. By subsidizing only conventional "end-of-the-pipe" treatment systems, the grant program induces planners to overlook what in some cases may be less costly or more effective alternatives, such as the storage (ponding) of wastes during critical periods of low-stream flow and the augmentation of the assimilative capacity of the stream through instream aeration and other devices.

2. Current regulations require secondary waste treatment for *all* municipalities along a watercourse. An optimal basin-wide plan would relate the degree of desired municipal treatment to steamflow conditions and downstream users (among other variables). In this optimal plan, some municipalities may require tertiary treatment while others may require only primary treatment. The drive for uniform secondary treatment results in excessive treatment costs at some outfalls and insufficient treatment costs at others.

3. States have failed to target federal funds on the municipalities with the most harmful discharges. More than one town situated

---

[6] Comptroller General of the United States, *Examination into the Effectiveness of the Construction Program for Abating, Controlling, and Preventing Water Pollution*, Washington, D.C., November 3, 1969.

downstream from major industrial locations has used federal funds to build treatment plants with the result that their treated effluent is of higher quality than the river into which it is discharged. Also, federal funds have been concentrated on smaller, largely suburban communities rather than on the larger cities with the most pollution. For example, nearly 40 percent of the federal grant money has gone to towns with populations of less than 10,000, and these communities contain less than 16 percent of the United States urban population. The largest cities, containing 25 percent of the total urban population, have received only 6 percent of the total federal grant money.[7]

4. The *construction* of treatment facilities does not guarantee their effective *operation*. In fact, the structure of the recent program creates incentives that work in the opposite direction. A second study by the General Accounting Office has confirmed the widely held belief that municipal plants are often operated inefficiently.[8] Over one-half of the plants surveyed were providing inadequate treatment due to overzealous efforts to reduce plant operating costs, the difficulty and expense of hiring trained personnel to operate the plants, and the failure of cities to maintain and repair equipment. By subsidizing only one part of the costs of effective waste treatment —plant construction costs—the federal government has induced resources into construction activity but has provided no similar inducement for efficient plant operation.

5. Federal grants for municipal waste treatment plant construction provide an indirect subsidy to industrial and commercial waste dischargers. By subsidizing the capital costs of municipal treatment facilities, the existing policy tends to reduce the sewer charges imposed on industrial, commercial, and domestic waste dischargers connected to the sewerage system. Because approximately 50 percent of the wastes handled by municipal treatment plants is from industrial sources, the size of the subsidy to business is substantial. The effect of this subsidy is to weaken the incentives for waste dischargers to seek alternatives to the public treatment of their waste flows. Production process changes, recycling, and materials recovery are all alter-

[7] Federal Water Pollution Control Administration, *The Economics of Clean Water*, Vol. I, Washington, D.C., 1970, p. 111.
[8] Comptroller General of the United States, *Need for Improved Operation and Maintenance of Municipal Waste Treatment Plants*, Washington, D.C., 1970.

natives to sending wastes to the municipal plant for treatment at public expense. These alternatives are numerous and are often less costly. Yet because these alternatives are not eligible for federal subsidies, firms will overlook them in favor of having the federal government pick up the tab. The tax subsidies have a similar distorting effect on the decisions made by firms as to the techniques they choose to reduce their discharges.

Thus, through federal grants for municipal waste treatment facilities as well as through tax subsidies for industrial pollution control equipment, current policy is, in effect, allowing polluters to generate and dispose of large quantities of wastes without bearing the full cost of their discharges—and then using taxpayers' money (at a current annual rate of about $1 billion) to clean up after them. With little effective constraint on the generation of industrial wastes and with rapid economic growth, the burden on the environment will skyrocket in the coming decades.

Turning to the regulatory-enforcement aspects of present policy, it is widely agreed that enforcement has not been effective so far, either at the federal level or at the state level where the primary responsibility now lies.[9] This is to be expected, as the discussion in Chapter 5 showed. The inherent difficulties with the regulation-enforcement process are nowhere more apparent than in the recent attempts to make the 1899 Refuse Act work. This law prohibits the discharge of "any refuse matter of any kind or description whatever" into any waters unless the discharger has obtained a permit from the U. S. Army Corps of Engineers. Virtually none of the present estimated 40,000 industrial dischargers hold valid permits.

This all but forgotten law made headlines beginning in 1969 as charges were brought against several firms as a consequence of individuals' initiatives. Some convictions were obtained, fines levied, and rewards paid to vigilant citizens as the law provides. But in June 1970, as the number of cases began to increase, the Justice Department attorneys were issued guidelines that instructed them not to bring charges against firms holding a permit issued by a state or local government. With this action, the Justice Department, in effect, established a policy of selective nonenforcement of one of the nation's laws. This policy was reenforced in December 1970, when the Administration announced that, while permits were required of all dischargers, including those "exempted" by the earlier Justice Department policy, no prosecutions

[9] For documentation, see David Zwick and Marcy Benstock, eds., *Water Wasteland, The Ralph Nader Study Group Report on Water Pollution*, New York: Grossman, 1971.

would take place as long as an application had been filed by July 1, 1971 and not subsequently rejected by the government. The effect of this is to grant firms an immunity from prosecution that is likely to last for several years as the government works its way through the backlog of unprocessed permit applications. And, since all permit applications are forwarded to the relevant state for certification that the discharge is consistent with the state's pollution control plan, the new policy has contributed nothing to the establishment of a more effective pollution control effort. It has, however, generated a flood of new permit applications to burden the workload of federal and state officials while, in effect, repealing the one potentially effective federal law against pollution.

NEW DIRECTIONS

This negative assessment of present pollution control efforts strongly argues for a major redirection of federal water pollution policy. Any new departure in policy must embody two features absent in present policy if it is to be successful. First, any new policy must recognize that a river is a system and that it cannot be efficiently managed in piecemeal fashion by numerous agencies and jurisdictions. It must emphasize comprehensive regional river basin planning and water quality management, and it must provide for the utilization of low flow regulation, in-stream reaeration, and regional collective treatment plants where these are appropriate. Second, any new policy must emphasize economic incentives— prices and charges—as a means of reducing the output of industrial and municipal wastes.

It is disappointing to contrast this prescription with the emerging legislative response to the growing awareness that present policies are not working. There are two developments worthy of note. The first is Senator Proxmire's bill to establish effluent charges at the national level. His bill would establish a system of uniform national effluent charges on industrial polluters. Half the revenues would be earmarked to the construction grant program for municipalities, and half would go to regional water quality management agencies to assist them in planning and carrying out their responsibilities.

On the positive side, this is the first proposal to incorporate economic incentives into our pollution control policy to be formally considered by Congress. The bill gives explicit attention to both of the essential features of any true reform of present policy, namely, incentives and a regional approach to planning and managing water quality.

On the negative side, the uniform national charge provided for in the bill would quite likely be unnecessarily high in some regions, leading to excessive investment in pollution control there, while being too low in other areas, leading to a failure to achieve water quality standards. One of the functions of an effective regional water quality management authority should be to establish a system of charges differentiated by location to take into account different water quality standards and assimilative capacities of rivers.

What is most discouraging, though, is the cool reception given to the Proxmire Bill in the Senate. Although it was first submitted in November 1969, it has yet to be reported out of Committee for consideration on the floor of the Senate. Instead, Senator Muskie's Subcommittee on Air and Water Pollution has responded to the growing pressure to do something about the failure of our present policy by drafting a bill that essentially does more of the same, only bigger and better.

The so-called "Muskie Bill," which passed the Senate by an 86-0 vote late in 1971, essentially ends the use of water quality standards as the measuring rod for performance and substitutes standards or regulations regarding effluent control and treatment. The bill requires that industry apply the "best practicable" waste treatment technology available by 1976, and by 1981 completely eliminate discharges of pollutants if it can be done at "reasonable cost." If zero discharge cannot be attained at reasonable cost, industry must install the "best available treatment" facilities, "taking into account the cost." A date of 1985 is set as a target for zero discharges from all sources of pollution.

The basic weakness of the bill is its reliance on precisely the sort of discretion that has led to the undermining of numerous other rule enforcement efforts. Who is going to define "reasonable," "best available," and "best practicable," and who is going to "take into account the cost?" All of these terms are open to widely divergent interpretations and none have precise legal meanings.

Another matter is the cost to the nation should the rules actually be enforced. Given existing technology and foreseeable developments, attainment of either the best available technology or the zero discharge goals will be prohibitively expensive in terms of national resources devoted to water pollution control. Because costs rise swiftly at very high treatment levels, the total cost of meeting these objectives could be several times that required to meet current quality standards and far in excess of any benefits resulting from elimination of the last few units of residual discharge.

The policy of subsidies to municipalities is also reaffirmed and extended in this bill. Appropriations of $20 billion are authorized through the fiscal year 1975, and a minimum of secondary waste treatment is required for all municipalities by 1976. The share of municipal waste treatment construction costs to be covered by federal grants is also enlarged.

Even while the Senate has pushed on with a bigger, more cumbersome, and thoroughly uneconomic version of the regulatory/ enforcement strategy, several states have given serious consideration to effluent charge legislation, and Vermont and Wisconsin have enacted a modified form of economic incentive for water pollution control.

The policy alternatives and the issues are becoming more clearly defined. The stakes are large, both in terms of the resources involved and the possible impact on the quality of life. But as we discuss in Chapter 9, there is some doubt as to the ability of a pluralist political system to make wise choices in issues of this sort.

7

Public Policy for Air Pollution Control

In this chapter, we focus on the problem of air pollution and the response of government to it. We view the air mantle as a commonly held resource yielding services of value to people, and we emphasize the need to manage this resource to maximize the social value of its services.

Before turning to the benefits and costs of air pollution control, it would be valuable to point out three ways in which the air pollution problem is different from, and perhaps more ominous than water pollution.

The most basic difference involves the sorts of damages or costs that the two kinds of pollution impose on people. In Chapter 6, we observed that polluted rivers and streams impose the bulk of their costs in the form of curtailed recreational activities—fishing, swimming, boating. Air pollution, on the other hand, imposes more direct costs on people. These costs consist of impairment of health, reduction of life expectancy, soiling of homes and clothes, impairment of visibility and of the growth of flowers, trees, and shrubs, and accelerated physical deterioration of public buildings and monuments.

A second basic difference between the two kinds of pollution concerns the number of people who are affected by it and the difficulty of escaping from it. In the case of water pollution, the people most heavily affected are those who seek opportunities for water-based recreation. While this may include most of us at one time or another, water pollution does not typically impose continuous and persistent inconvenience and cost on us. The case of air pollution is different, however. Airborne residuals are emitted into the atmosphere of the cities in which most of us live and breathe. The people affected are most of us all of the time. Consequently, the costs of avoiding the effects of pollution are enormous. For most city residents, avoiding air pollution would entail the sacrifice of current jobs and homes and the change in physical location from urban to rural areas. These costs loom substantially larger than those confronted by recreationists in searching for alternatives for natural water bodies.

A final distinction between the two forms of pollution concerns the relative difficulties confronted by government in controlling and reducing their adverse effects. To discharge residuals into a river, one has to have access to the river; residuals can be placed into the air from any geographical point. Hence, detecting and monitoring sources of emissions is more difficult for air pollution than is the case with water pollution. Reinforcing this difficulty (and related to it) is the fact that water-borne residuals can be and are collected in sewer systems and can be and are brought to a central facility for treatment. It is inconceivable to think of the collection and central treatment of the residuals spewed into the air of a city. Finally, it is possible, and in some cases economically feasible, to augment the assimilative capacity of rivers. There is no known technology for augmenting the capacity of the air mantle.

THE BENEFITS AND COSTS OF IMPROVING AIR QUALITY

Throughout our discussion we have emphasized that the formulation of efficient environmental management policy must rely on the evaluation and comparison of the benefits and costs of improvements in levels of environmental quality. As in the case of water pollution, it is a good deal easier to identify the components of benefits and costs of improving air quality than it is to measure the dollar values of these effects.

As with water, the benefits are the damages avoided when air quality is improved. There are two types of damages—those to property and those to humans, either in the form of impaired health and shorter life or in the loss of utility of amenity associated with enduring dirty air. Ideally, we would like to observe how people respond to differences in air quality in order to infer something about their willingness to pay to avoid dirty air or to obtain the benefits associated with cleaner air. This approach is made much more difficult in the case of air quality because the differences in air quality and in the effects of polluted air may or may not be perceived by individuals, or they may lack knowledge of cause-and-effect relationships. For example, although they may perceive air pollution in the form of particulate matter settling on their cars, they may not be aware of the effect of particulate matter on their lungs and on their life expectancy. Similarly, they may perceive certain phenomena, such as vegetation damage, and not know that the cause was an air pollutant.

Concerning the effects of air pollution on human health, laboratory and clinical work has helped to establish some of the causal links between certain polluting substances and observed effects on human health. But this is insufficient for estimating health damages or benefits. First, it is necessary to estimate statistically the relationship between concentrations of air pollutants and mortality rates and the incidence of ill health. Comparisons among cities, regions, or localities are usually the source of data for this kind of study. One of the principal empirical difficulties is that the air quality data usually refer only to a single point within the geographic unit under study, and air quality may be quite different in other parts of the city or region. Furthermore, the mobility of people among cities may mean that a person who is recorded as dying of lung cancer in Phoenix may have lived all but the last six months of his life in Pittsburgh.

These difficulties notwithstanding, researchers have uncovered a number of statistically significant relationships between various measures of air quality and mortality rates for different causes.[1] From these relationships it may be possible to estimate the change in death rates and change in life expectancy associated with a specified improvement in air quality. The next question is the value to be placed on the increase in health and/or life expectancy. It is clear that people do make decisions and choices that involve choosing a slightly longer or shorter life expectancy in return for a little less or more of something else. Some people take airplanes when they could ride trains, but others do not fly. It is very difficult, however, to move from this common-sense observation that people do place values on changes in their life expectancy to meaningful estimates of what the values are.

Researchers have found it necessary to make some simple but arbitrary assumptions to cut through this problem. The first assumption is that if a premature death because of air pollution is prevented, the value of the additional life is the amount of money that the person will earn during his remaining life. This equation obviously understates the value of life in one way, because it

[1] See Lester B. Lave and Eugene P. Seskin, "Air Pollution and Human Health," *Science,* 169 No. 3947 (August 21, 1970), 723–733; Lester B. Lave, "Air Pollution Damage: Some Difficulties in Estimating the Value of Abatment," in Allen V. Kneese, and Blair T. Bower, ed., *Environmental Quality Analysis: Theory & Method in the Social Sciences,* Baltimore: The Johns Hopkins Press, 1972; and William Hickey, "Air Pollution" in William Murdock, ed., *Environment: Resources, Pollution, and Society,* Stamford: Sinauer, 1972.

places a zero value on the life of nonworking persons and the elderly. On the other hand, one could question whether an individual would be either willing or able to give up *all* his potential future earnings in order to prolong his life. The second assumption is that if a person is sick, the cost to society is the income he would have earned if he were healthy plus the hospital and doctors' costs and the like associated with treating his illness.

Using statistical estimates of the relationship between air pollution levels and death rates, a recent study has estimated the reduction in deaths that can be expected to accompany a 50 percent reduction in particulate and sulfur oxides concentrations across the nation.[2] Then, using the assumptions described above to place economic values on health and life, the study finds that as of 1963, the postulated reduction in air pollution would be worth about $2 billion per year.

While ill health is the most important damage associated with air pollution, others occur as well. Because air pollution corrodes metals and damages stone and masonry, people would benefit from improved air quality by having to incur less expense for maintaining, repairing, and replacing buildings and other structures. Similarly, improved air quality would eliminate much of the damage that pollutants now inflict on crops, trees, and other vegetation. These damage reductions must also be calculated as benefits attributable to improved air quality.

The final major benefit from effective air pollution control stems from the increased value of land and property sites near central cities and other highly polluted areas. Although many of these sites have the attractions of being closer to central business districts and easily accessible by mass transportation, they are often viewed as unpleasant places to live because of high levels of air pollution. Improvements in air quality in these locations would increase the demand for them and as a result increase their price. In a real sense, these improvements in property values are also benefits attributable to improved air quality. However, to some extent, these may duplicate such benefits as less cleaning and corrosion, since people may move away from areas where these effects are noticeable and thus depress the real estate market. There is probably little double counting with health benefits since people usually seem to be unaware that their health is being impaired.

After surveying all of these sources of benefits from improved air quality, the President's Council on Environmental Quality con-

---

[2] Lave and Seskin, op. cit.

cluded that the elimination of air pollution would confer benefits on people of over $16 billion per year. They stated:

"The annual toll of air pollution on health, vegetation, materials, and property values has been estimated . . . at more than $16 billion annually—over $80 for each person in the United States. In all probability, the estimates of cost will be even higher when the impact on esthetic and other values are calculated, when the cost of discomfort from illness is considered, and when damage can be more precisely traced to pollutants."[3]

To attain these benefits from improved air quality will require changes in the patterns and techniques of both consumption and production activities. As in the case of water pollution, these changes do not come easily, nor are they costless. The development of new technologies, the substitution of low for high-residual production processes, and the treatment of emissions prior to their release into the atmosphere all require the use of resources. These costs of improved air quality cannot be ignored in determining appropriate air pollution control policy.

Documenting the costs of controlling air pollution is almost as difficult as evaluating the benefits. This is so in large part because the techniques that will be chosen to reduce the residuals discharged into the atmosphere are not known with any precision. Indeed, in some cases the technology is not yet developed. For example, the quantity of sulfur oxides released by a power-generating plant can be reduced by using higher-priced low-sulfur coal, by improving the efficiency of the boiler, by running the emissions through a curtain of water (called scrubbing), or by producing less electricity. The costs of reducing the volume of pollutants will be different in each case.

In 1971, the President's Council on Environmental Quality reported its estimates of the costs required to meet already established air quality standards. These are presented in Table 7-1. They estimated that for the six-year period 1970 to 1975, nearly $24 billion would have to be spent to achieve these standards. This compares with an estimate of nearly $40 billion to achieve water quality standards. In both cases, this assumes continuation of the relatively inefficient strategies we are now following.

Of this $24 billion of estimated air pollution control costs,

[3] See President's Council on Environmental Quality, *Environmental Quality,* 1971, 2nd Annual Report, p. 107. This estimate is based on a review of other studies, such as Lave and Seskin's, and probably should be considered to have a margin of error of plus or minus 50 percent.

Table 7-1   The Costs of Air Pollution Control
(in Billions of Dollars)

| | Capital Investment | Cumulative Costs for Meeting Standards between 1970–1975 | |
		Total Operating Costs	Total Expenditures
Federal government facilities	0.4	1.2	1.6
Private			
Mobile sources	5.4	0.6	6.0
Stationary sources	8.0	8.1	16.1
Total	13.8	9.9	23.7

Source. See President's Council on Environmental Quality, *Environmental Quality—1971*, Washington, D.C., p. 111.

about $1.6 billion would be spent by the federal government to control pollution from its own facilities; $6 billion would be in the private sector to reduce automobile and other vehicle emissions, and $16 billion would be privately incurred costs to reduce the emissions from manufacturing plants, electric utilities, and other stationary sources. Clearly, the bulk of these private sector costs will be borne by consumers in the form of higher prices for the commodities that they buy. One of the commodities that will be the hardest hit by the air quality standards already legislated will be automobiles. For example, it is estimated that if automobile emission devices are used to achieve the emission standards established by the latest legislation, car prices will be increased by between $250 and $400 by 1975.

Although some of these costs appear staggering, it must be recalled that estimates of the benefits of improved air quality are also huge. As we emphasized in the discussion of water quality improvement, public policy must strive to strike the balance between too little and too much improvement in air quality. That balance is struck when the last $1 expended to improve air quality is just matched by $1 of benefits from reduced air pollution—provided that the improvements in air quality are achieved at the lowest possible cost. But, as with water pollution, in practice, air quality standards will be set on the basis of some rough weighing of benefits and costs in a political context. The economic problem then becomes one of achieving the standards at least cost.

THE PRESENT FEDERAL POLICY

Federal policy to abate air pollution has adopted a regulatory/ enforcement strategy similar to that of federal water pollution control policy. However, because airborne wastes can not be collected and treated in one central facility, there has been no subsidy comparable to the construction grant program for water pollution. Nor are regional collective facilities likely to play a role in optimal control policies. But again, economic incentives have not been recognized as the powerful public policy instruments they potentially are to achieve improved air quality.

The federal government first showed concern with the problem of air pollution in the mid-1950s. It was at about this same time that federal action to improve water quality also began. The first air quality legislation was passed in 1955. That law authorized modest appropriations for research, data collection, and technical assistance. No provision was made for the establishment or enforcement of air quality standards or other control measures. The early law was supplemented in 1960 when the U.S. Public Health Service was authorized to study the effects of motor vehicle exhausts on health. The Clean Air Act, passed in 1963, expanded the research effort devoted to motor vehicle exhausts and fossil fuel combustion, authorized grants to state and local agencies in support of the development of air pollution control programs, and permitted direct federal action to abate *interstate* air pollution. However, as in the case of the 1956 Water Pollution Control Act, court action to enforce pollution control could be taken only after a series of hearings and conferences had been held.

The policy direction begun by the 1963 Act was reinforced and broadened in both 1965 and 1967. In 1965, Congress authorized the establishment of national standards for motor vehicle pollution which were to be applied to 1968 model vehicles. The Air Quality Act of 1967 focused in "air sheds" and air quality rather than individual pollution sources, and sought to establish control programs for achieving air quality standards. The Act required the Department of Health, Education, and Welfare (HEW) both to delineate relevant "air sheds" and to develop and publish air quality "criteria" which describe the effects of each of the major pollutants on health, vegetation, and physical materials. The law also required HEW to publish descriptions of the techniques available for abating air pollution together with their costs and effectiveness. For the first time, the 1967 law required states to establish air quality standards in each of the designated air sheds. These standards were to prescribe maximum concentrations of the pollutants described in the criteria documents. After the standards were set, states were to develop plans for achieving the standards. These plans would set specific emission levels on individual sources and lay out timetables for achieving compliance with the emission levels set. However, this program never got off the ground. The federal government was slow to establish regions and publish criteria documents, and without them, states were unable to set standards.[4]

The breakdown of the procedure adopted in 1967 led to the passage late in 1970 of a most comprehensive and sweeping piece of legislation—the Clean Air Amendments of 1970, sometimes known as "the Muskie Clean Air Bill." First this law set deadlines for the delineation of air quality control regions for setting ambient air quality standards by the states, and for preparing implementation plans for achieving these standards. These deadlines have now passed, and all states have by now submitted air quality standards and implementation plans for the federal review and approval provided for in the law. The adequacy of the standards and implementation plans submitted by some states is seriously in question, and in these cases, actual implementation will be delayed until the federal government and the state can reach agreement on the points at issue.

In a major new departure from past practice, the 1970 law supplemented enforcement of ambient air quality standards by the states with a system of *national emission standards* to be applied to new stationary sources of air pollution. This feature empowers the Administrator of the EPA to establish maximum allowable rates of discharges for different classes of sources, for example, pulp mills and electric power generators, and to enforce them through the federal courts. These emission standards would be independent of air quality standards in particular regions. Their purpose is to assure that new facilities incorporate the best available control technologies.

The most widely publicized feature of this act was its establishment of stringent limitations on emissions from automobiles. The law requires that carbon monoxide and hydrocarbon emissions be reduced by 90 percent from the standards in effect in 1970, and that nitrogen oxides emissions be reduced by 90 percent from the uncontrolled levels of 1970. In addition, the law requires that automobile manufacturers give customers a 50,000 mile or five year warranty that the vehicle will continue to meet the standards. Since the automobile poses some rather difficult problems for air pollution control policy makers, we will devote a separate section

---

[4] See John C. Esposito, *Vanishing Air: The Ralph Nader Task Force on Air Pollution,* New York: Grossman, 1970.

to raising some of these problems, evaluating present policies, and discussing some alternatives.

THE AUTOMOBILE—A SPECIAL PROBLEM

The auto has a highly developed capacity to inflict external costs on others than its own driver. It is not only air pollution, but the congestion of highways and city streets, the blight of highway construction, and the threat of sudden and violent death from the other lane which make the auto, at best, a very mixed blessing, or as some would have it, an unmitigated disaster. In this section we deal only with some aspects of the air pollution problem. We show that even from this narrow perspective, there are no easy answers, and we can only hint at the difficulty in dealing with the auto and all of its ramifications as a major social problem.

Which Pollutants are Most Important?

Autos are major sources of three important air pollutants: carbon monoxide, hydrocarbons, and nitrogen oxides. Which of these causes the most damage and should be most controlled? We do not know. The economist's logic would have us develop the control technologies for the most dangerous substance first, and establish the strictest standards on this substance. Yet we have adopted essentially the same standards for all three substances (90 percent control). At least until the newest federal law, the standards that were adopted were dictated by technical feasibility, rather than the effects of the pollution on people.

In other words, contrary to the economist's logic, resources have been allocated to controlling the three pollutants solely on the basis of the costs of control with no attention being given to the potential benefits. One consequence has been that there will be no controls on nitrogen oxides before 1973. No one would seriously argue that this is the least dangerous of the pollutants; instead, a case can be made that nitrogen oxides pose the most serious threat to human health.

Maintenance—Whose Responsibility?

At present, emission standards are imposed on the manufacturer. Cars are tested before they are sold, but there is no follow-up to assure that cars continue to meet the standards. A number of studies have shown that a high percentage of cars on the road cannot meet the standards set for them after as little as 10,000 to 15,000 miles of use. Auto maintenance and good tuning of the engine are at least as important in limiting emissions as the design and installation of control equipment. The 1970 law now requires manufacturers to guarantee that their cars will continue to meet the standards for 50,000 miles or five years. There are two problems here. First, enforcing the warranty provision is likely to require a cumbersome and costly administrative system. Second, manufacturers are being pushed in the direction of designing a maintenance-free engine system in order to satisfy the warranty requirement; this may be quite costly, at least in comparison with relevant alternatives.

The problem with the present control system and standards is how to get the car owner to tune and maintain his engine periodically. Almost twenty years ago economists at Rand Corporation proposed the answer—a Smog Tax.[5] In one version of this tax, cars would periodically be tested and assigned a smog rating. The rating could be indicated by a seal or coded device attached to the car. Then, when the driver purchases gasoline, he would pay a tax over and above the present gasoline taxes, and the amount of the tax would depend on his smog rating.

Under the smog tax system, an individual can reduce his smog tax bill in several ways.

1. Tuning up or overhauling his engine to reduce emissions and obtain better gas mileage. This can be a highly effective alternative. New Jersey recently established emission standards for cars registered in the state with the standard being less stringent for earlier model years. In a pilot study, they found that about half the cars tested failed the state standards. But more important in the present connection, they found that in almost every case of failure, cars could be made to pass the test by undergoing a regular engine tune-up by a competent machanic. The cost per car averaged about $20.[6] The New Jersey work shows that vehicles can be efficiently tested (it takes about 35 seconds) and that engine condition, including recency and quality of tune-up, are extremely important in terms of what is actually emitted.

[5] D. M. Fort, et al., "Proposal for a Smog Tax," Reprinted in U.S. House of Representatives, Committee on Ways and Means, *Hearings—Tax Recommendations of the President,* 91st Congress, 2nd Session, September 1970, pp. 369–379.

[6] New Jersey Department of Environmental Protection, Bureau of Air Pollution Control, "Motor Vehicle Tune-up at Idle" (no date) and "Notice of Public Hearing," May 27, 1971.

2. Driving fewer miles per year. The car owner has many options here. He can move closer to his job, use mass transit, and enter car pools. Standards based on emissions per vehicle-mile do nothing about miles driven. The smog tax affects vehicle-miles as well as emissions per mile.

3. Purchase control devices for older cars. On June 29, 1970, General Motors announced failure of a test campaign to market control kits for pre-1968 models. These kits cost about $20 installed, but not one would buy them. Clearly, it is nonsensical to think people would, since, if there is no assurance that other people will buy them, there is no incentive for any single individual to do so.

4. Buying a car with a better smog rating. Since there would be a demand for them, manufacturers would have an incentive to design automobiles with better smog ratings not only at time of manufacture but throughout their life.

Only the first of these four alternatives is relevant to the question of who should be responsible for the continued attainment of emission standards. But the New Jersey study suggests that placing the responsibility on owners and backing it up with appropriate economic incentives is practical. Not only that, but the tax elicits other desirable responses from drivers, such as points 2–4, and its incentives apply to owners of pre-1968 cars that have no control systems, but whose emissions can often be cut substantially by better maintenance.

Los Angeles versus Watertown, South Dakota

Los Angeles has a smog problem. Watertown, South Dakota does not. Why should the same emission standards apply to cars in both locations? There is no really good reason. National emission standards will force millions of car owners in rural and less densely settled areas to pay part of the $5.4 billion cost of controlling auto emissions between 1970 and 1975, but will provide little or no benefit to them. Their expenditure is the price of having a system that is easy to administer at the federal level.

Is there an alternative? Again, the smog tax has some attractive features. The smog tax rate could be very high in Los Angeles and low or even zero in areas with no auto air pollution problem. Then people could buy the most expensive low-emission cars if they expected to drive a great deal in high-tax areas. But rural residents would not be forced to incur costs without benefits.

Getting the Lead Out

Tetraethyl lead is added to gasoline because it is presently the cheapest way to raise its octane rating, but continuing to use lead in this way raises two kinds of problems. First, although the evidence is not conclusive, it appears that lead itself is an air pollutant and is or perhaps will soon be imposing external costs on people.[7] The second problem stems from the auto industry's need to curb emissions of carbon monoxide, hydrocarbons, and nitrogen oxides. It appears that the auto industry is planning to use catalytic reactors in the exhaust pipe to complete the combustion of the fuel, converting CO and hydrocarbons to CO_2 and water. They may also choose to use a different kind of catalytic reactor to convert nitrogen oxides back to nitrogen and oxygen. In either case, lead in the gasoline fouls the catalytic reactors, rendering them useless after a few thousand miles of driving. Thus the auto industry argues that eliminating lead additives from gasoline is essential to their efforts to meet the 1975 emission standards.

Of course, the oil companies do not want to stop using lead, since they would incur substantial expense in modifying their refineries to produce gasoline of adequate octane without lead. But this is not all there is to the story. Unleaded gasoline of equivalent octane contains more of the volatile aromatic fractions. This means that the gasoline evaporates faster and the hydrocarbon vapors are more photochemically reactive; they make smog. Thus, less lead may mean more smog.

Finally, there is an alternative technique for controlling emissions available to the auto industry. It is a thermal reactor; it is not adversely affected by leaded gasolines, but it can only be used in conjunction with an extensive, and expensive, redesign of the front end of the auto. Using catalytic reactors involves no major redesign or modification of the car or engine. Therefore the issue of lead in gasoline can be viewed in part as a battle between the oil and auto industries over who is going to bear the major burden of meeting the 1975 emission standards.

Because there may be a substantial difference in the total cost to society of achieving the standards by these two alternatives, and because lead, itself, is an environmental threat, the public has a substantial interest in the decision as to which technique will be used to control emissions. The decision will, however, be made, if indeed it has not already been made, in the boardrooms of the Big Three in Detroit. This is one instance where the firm's freedom to choose *how* it will meet established environmental standards is not an unmixed blessing.

[7] See Chapter 3, pp. 47–48.

A Broader View

Perhaps the full complexity of the automotive pollution problem can better be seen as we attempt a broader perspective. We have suggested that a smog tax would be useful in dealing with two aspects of the problem, maintenance of control systems on older cars and different regional situations. We believe that such a tax would have a substantial impact on automotive emissions in the short run by eliciting a variety of other responses from drivers. But it is not so clear that a smog tax would impel the auto industry toward a "correct" choice as to how to meet the coming 1975 standards. Nor is it clear that the smog tax will be much help as we consider alternatives to the internal combustion engine, as soon we must. An attractive feature of an economic incentives strategy is that it elicits graduated responses that are related to marginal benefits and costs. But, for example, the development of major mass transit systems as an alternative to continued reliance on the automobile requires the kind of large-scale mobilization of resources that is beyond the scope of markets and market-related pollution control strategies.

THE SULFUR OXIDES TAX

In 1971, the President proposed a tax on sulfur oxides emissions. This tax would be levied on electric power plants that burn coal and oil to generate electricity, refineries, smelters, and perhaps other industrial processes. Among the primary pollutants, sulfur oxides are one of the most damaging, being linked to both lung cancer and bronchitis, as well as damage to vegetation. To date, reduced emissions of sulfur oxides have come primarily from switching from high-sulfur to low-sulfur fuels. The technologies for removing sulfur from fuels before burning or from the exhaust streams are on the drawing boards but the past five years have seen little movement toward development of these alternatives. This seems to be a case ideally suited to an economic incentives strategy.

Levying a per pound charge on the sulfur emitted by power plants and other industrial firms would provide economic incentives to undertake a variety of economic responses resulting in improved air quality. These can be described as follows:

1. The potential cost savings would provide a strong incentive for the development and installation of effective sulfur removal systems.

2. Users of coal and oil are confronted with a range of fuel options, some with high sulfur content, others with low sulfur content. Currently, the low-sulfur fuels are more expensive. The charge would provide the incentive required for fuel purchasers to choose the low-sulfur fuels voluntarily.

3. Oil refiners now have little incentive to remove the sulfur from fuel oils in the refining process. The sulfur oxide emission charge, by creating a demand for low-sulfur fuels, would provide refiners with a strong incentive to remove sulfur in the refining process and to develop new techniques for doing it at lower cost. The sulfur content of coal can also be reduced by processing.

4. By raising the price of commodities whose production processes employ coal and oil for combustion, the emission charge would tend to reduce the quantity demanded of these commodities. As a result, consumers would tend to shift their pattern of purchases to commodities with less serious environmental effects.

The example of the sulfur tax also gives us an opportunity to point out that the details of a plan and consideration of possible side effects may be as important as adoption of the principle of residuals charges.

When the details of President Nixon's proposal were finally made public in February 1972, it was found that the charge would be imposed only in those air quality control regions where the ambient concentrations of sulfur oxides exceeded federal standards. In these regions the charge would provide an incentive to reduce emissions sufficiently to achieve the standards and eliminate the charge. The charge would also provide an incentive for firms to move operations from "dirty" regions to clean regions to avoid paying a charge, and new plants would most certainly be located in regions of high air quality in preference to dirty regions where a tax would be imposed. Therefore over a longer period of time, shifts in industrial location would degrade the quality of air in the cleaner regions and bring the country down to the lowest common denominator as represented by the federal standard. How can this be prevented? Impose a charge in all regions.[8]

The materials balance principle reminds us of a second side effect. If sulfur is removed from the flow of residuals gases, it must go somewhere. Actually this need not be a problem since raw sulfur is a valuable commodity. The hitch comes when we look

[8] But as we pointed out in Chapter 6, provision for different charges in different regions is preferable to a single uniform nationwide charge.

at the magnitudes involved. The present production of raw sulfur in the United States is around 10 million tons per year. A 50 percent recovery of the sulfur being dumped into the atmosphere would yield an additional 8 million tons per year. Any such increase in supply would seriously disrupt the sulfur-producing industry, the employees it hires, and the regions in which it is located. Before finally implementing such a charge, equitable and efficient means of dealing with such dislocations must be considered. Because of such market linkages, planning for an economic approach to environmental control requires consideration of labor retraining, migration, and regional adjustment assistance issues.[9]

CONCLUSION

In general, the federal response to the challenge of air pollution has been characterized by an excessive reliance on enforcement of emission standards imposed on both stationary and mobile sources, and therefore is likely to lead to diminished effectiveness and higher than necessary costs for cleaner air. There are promising opportunities for adopting economic incentives, as the smog and sulfur tax proposals indicate. These must, however, be carefully designed to avoid creating perverse incentives (as in the Nixon version of the sulfur oxides tax). The potential benefits from controlling air pollution appear to be large, and while the estimate of the costs of achieving air quality objectives are also large, it appears that a more economic approach to air pollution control could yield substantial savings.

[9] Approaches to these problems are discussed in Chapter 8. See pp. 147–148.

8

Environmental Management: Some Issues

THE COSTS OF POLLUTION CONTROL

The discussion in the last two chapters shows that pollution control and environmental protection cannot be obtained without incurring significant costs. No one can state with certainty what the final bill for the whole economy will be. This is true partly because of the difficulties inherent in estimating and predicting on this scale and partly because we are still in the process of making some of the choices and decisions that will influence the size of the bill. The total costs will ultimately depend on three factors: how much environmental improvement we want; the state of the technology of pollution control and the rate of improvement in this technology over time; and the nature of the pollution control policies chosen to implement our environmental goals. As we have shown, some pollution control policies are better than others in terms of their ability to achieve pollution control objectives at least possible total cost.

The President's Council on Environmental quality has presented its estimates of the total costs for the six-year period 1970 to 1975. We have already referred to some of the individual components of them in earlier chapters. The totals are:[1]

For water pollution control	$38 billion
For air pollution control	23.7 billion
For solid waste control	43.5 billion
Total	$105.2 billion

---

[1] The President's Council on Environmental Quality, *Environmental Quality—1971*, Washington, D.C., 1971, p. 111.

Most of the solid waste figure represents continued expenditures for existing collection and disposal services. If we omit these, we obtain an estimate of the *incremental* expenditure for air and water—a total of $61.7 billion. A recently published survey by McGraw-Hill, Inc. reveals what industry estimates it will have to invest in new pollution control equipment by 1976. Industry estimates that it will have to spend $22.8 billion through 1975. Planned expenditures for 1972 are $4.9 billion. An average annual spending of $5.7 billion per year over this period will be needed to meet their estimated total requirement.[2] In contrast, the Council on Environmental Quality estimates industry's capital requirements to be $18.7 billion; but in addition they foresee spending of $16.9 billion for annual operation.[3]

Both the Council on Environmental Quality's and McGraw-Hill's estimates of spending needs are rough approximations at best. We do not yet know enough about the costs of carrying out the pollution control policies we are in the process of developing. Yet, we can see enough of the picture to make several important observations. The first concerns the magnitude of these costs and our ability to pay the bill. Although the figures look large in themselves, they are relatively small in comparison with other relevant economic magnitudes. For example, the $61.7 billion total for six years for air and water, which is the highest official estimate yet published, is only about 6 percent of this year's gross national product and represents less than 25 percent of the *increase* in GNP that we can reasonably expect to occur because of economic growth over this period. Even the $105 billion figure which includes solid wastes comes to only about 40 percent of the expected economic growth "dividend." The question is not one of ability to pay, but rather one of priorities and *willingness* to redirect our economic resources from other things to pollution control.

Turning to industry's own estimate of required expenditures, pollution control spending is expected to be only 5.3 percent of total planned capital expenditures between now and 1975. This means that a doubling in the level of spending for pollution control by industry could be accommodated with only a 5.6 percent reduction in other capital spending. It is not that industry lacks the financial resources. Rather they lack an incentive to use more of these resources for pollution control.

While trying to put these large sounding totals in some kind of perspective, we do not want to minimize the political and economic difficulties in achieving a reallocation of resources of

[2] *Business Week*, May 13, 1972, p. 77.
[3] *Environmental Quality—1971*, op. cit.

this magnitude. As resources are moved out of one sector of the economy, for example defense and aerospace, and into the pollution control equipment industry, there are likely to be bottlenecks, shortages, and so on. While a market economy with its resource mobility, price signals, and decentralized decision making is remarkably well-suited for bringing about the necessary economic adjustments, adjustment can never be instantaneous. One consequence may be temporary unemployment.[4] There are likely to be price increases as well. But there is no reason to expect either persistent depression and unemployment or a continuously rising general price level. These are problems of macroeconomic stabilization. And without trying to minimize the difficulties in achieving and maintaining full employment without inflation, we can say that an effective pollution control program creates no more serious obstacles to these twin objectives than a host of other regularly operating economic forces.

Our second observation concerns the benefits of good planning. Although the total (cost) figure may be small relative to the size of the economy, it is large enough in magnitude to provide high payoffs for efforts to economize in pollution control. A change in policy that would reduce the total costs of pollution control by 10 percent would result in a substantial savings of money over time running to many billions of dollars. Every effort should be made to develop policies that will achieve pollution control goals at least possible cost and that will induce the kind of technological change that would make pollution control less expensive.

Third, there are good reasons for believing that pollution control goals can be met at less than the estimated costs if improved management approaches, including effective economic incentives, are adopted. We have developed the details of this assertion in earlier chapters but it deserves reemphasis. The available estimates focus on treatment of residuals streams already generated at the expense of other technologies that would enter into an optimum program for achieving environmental objectives. In the case of air pollution, for example, the emphasis is on treating gaseous residuals after generation—for example, removing substances from stack gases and providing "add-on" devices for internal combustion engines. Alternatives, such as fuel preparation, short-term fuel substitution during severe pollution episodes, switching to fuels having inherently low emissions, and substituting alternatives like steam or turbine engines for internal combustion auto engines are given less emphasis.

---

[4] This possibility is discussed in more detail in the next section. See pp. 144–146.

In the case of water, the neglect of alternatives to conventional approaches seems even more serious—perhaps because we know more about the range of potential choices. For example, the Delaware Estuary study (see Chapter 6) showed that the use of even a relatively simple system of economic incentives, in the form of effluent charges, could cut in half the social cost of achieving water quality standards in that river in comparison with a conventional approach like uniform treatment. The saving in cost over a 25 year period was about $150 million. Furthermore, careful studies of a number of basins, including the Delaware, Miami, Potomac, Wisconsin, Raritan Bay, and others, reveal that a regional management approach, including such technologies as mechanical reaeration of watercourses and reservoir storage to regulate low flows operated in a manner closely articulated with waste water treatment operations, can greatly increase the effectiveness and efficiency of water quality improvement programs.[5]

Fourth, studies of a number of industries—including beet sugar, petroleum refining, canning, pulp and paper, and wool reprocessing—show that process changes and changes in the mix of inputs and outputs can often be less costly ways of reducing industrial wastes than treatment after the wastes have been generated. Any program aimed at achieving efficiency in the pursuit of environmental quality must take account of these facts.

Finally, these estimates represent a lot of capital investment for catching up and correcting past abuses. Probably more than three quarters of the expenditures would be investment. Accordingly, expenditures discussed above are likely to "hump" and then drop off considerably. One should not make too much of this point, however, because some further steps could turn out to be extremely costly. For example, it has been estimated that it might cost $90 billion to remove or recycle nutrients from all treated sewage effluents to protect streams and lakes from excessive enrichment and eutrophication. Separation of storm and sanitary sewers in the major cities to prevent the overloading of treatment plants after rainstorms could cost as much as $50 billion. And last there are no well-founded estimates of the costs of controlling nonpoint-source water pollution.

In summary, the estimates discussed here suggest that while achieving major environmental improvements is likely to be very costly, the amounts are not so great as to require a significant reduction in our material well-being. Nor is unemployment or inflation inevitable. However, this is a short-run conclusion, valid

[5] See Allen V. Kneese and Blair T. Bower, *Managing Water Quality: Economics, Technology, Institutions,* Baltimore: The Johns Hopkins Press, 1968.

for perhaps a few decades. The implications of very long continued growth in population and material and energy conversion could be quite different. We offer some observations on this in a later section.[6]

EQUITY AND POLLUTION CONTROL

Crudely put, the equity questions are: Who gets hurt by pollution, the rich, the poor, or everybody? And who will wind up paying the costs of controlling pollution? Given our failure to come to grips effectively with the general national problem of equity in income distribution, there is good reason to be concerned with the distributional impact of particular pollution problems and policies.

Benefits

With respect to who gets hurt by pollution and who would benefit from effective pollution control programs, the situation seems to be roughly as follows. The poor, and blacks in particular, are exposed to polluted air more than affluent whites.[7] One of the advantages of wealth is that it enables its possessors to buy protection from environmental insults such as air pollution. The wealthy can live in the best suburbs, drive in air conditioned cars, etc. Accordingly, improved air quality in center cities would directly affect the poor and blacks more than the rich.

However, the ultimate impact of improved air quality on the distribution of income and welfare is more complicated. Economic reasoning suggests that improved air quality in a part of an urban area will make the land under that air more valuable and raise land prices. In fact, as we pointed out in the previous chapter, investigators have found that air quality and land prices tend to be related in urban areas, with higher land values in areas of higher air quality, other things being equal. If the urban poor and blacks tend to rent their dwellings rather than buy them (which is true), it may turn out that in the long run some part of the benefits of air quality improvement will be passed on to the landlord-

---

[6] See pp. 156–161.

[7] This finding and other points in this section are based primarily on A. Myrick Freeman III, "The Distribution of Environmental Quality," in Allen V. Kneese and Blair T. Bower, eds., *Environmental Quality Analysis: Theory and Method in the Social Sciences*, Baltimore: The Johns Hopkins Press, 1972.

property owner in the form of higher rents. In other words, while urban dwellers will be better off because of better health and reduced damages and amenity losses, this gain will tend to be partially offset by higher costs of housing. Nevertheless, it appears, on balance, that the improvement would be pro poor.

Turning to water pollution control, the principal readily identifiable benefits are in the form of improved recreational opportunities and amenities. People with higher incomes tend to participate in the various forms of water-based recreation more often than those from lower income groups; and they tend to place a higher dollar value on amenities, largely because they have more dollars. Thus, it appears that water quality improvements would tend to accrue disproportionately to the more affluent. This tendency is modified somewhat, however, by the fact that major improvements in water quality would often be centered in or near large cities, thus providing more opportunities for water-based recreation close to where poorer people are concentrated.

Costs

First we must distinguish between two broad types of pollution control costs—real *resource costs* and *factor income costs*. The resource costs represent the land, labor, and capital, that must be devoted to altering production processes, to recycling and recovering materials so as to reduce the amounts of pollutants generated, and to collecting and treating wastes that would otherwise be discharged to the environment. The factor income costs represent those changes in labor and capital incomes due to pollution control measures.

Consider first the real resource costs. Where public agencies undertake the pollution control activity (for instance, municipal sewerage treatment), the resource costs are passed on to the taxpayer, or perhaps are covered by user charges levied on those who transmit their wastes to the plant. Where user charges are imposed on firms, and where firms undertake pollution control efforts of their own, the resource costs will be passed on to consumers, for the most part, in the form of higher prices for goods and services.

The resource costs are all real costs or opportunity costs because they represent opportunities foregone. All the resources used in pollution control could have been put to some other use. The cost to society is what their value would have been in their next best use. On the other hand, factor income costs are not necessarily real costs. Pollution control programs may result in changes in factor prices as some resources become less valuable, and their owners may thus experience a reduction in income. For example, an air pollution control program may result in a lower price for high sulfur content coal. The coal may still be mined and used, thus there is no real or opportunity cost; but coal mine owners will receive lower incomes on their holdings of land containing coal.

Factor income costs can, however, be real costs— especially in the short run. For example, as a result of a pollution control policy that raises the real costs to firms, output will have to be contracted as the price is increased to cover the higher costs. This follows from the law of demand. With lower total output some plants, and perhaps some firms, will have to reduce output and perhaps go out of business. If the released resources that have alternative uses, especially labor, are quickly employed in other activities at comparable wages, there are no factor income costs. But if the resources remain unemployed for some time, because of geographic immobility, or imperfect information on job alternatives, there is a significant factor income cost. Moreover, this cost is a real cost, since the unemployed resources could be producing something of value for society.

What is the likely incidence of these costs, that is, who pays? Those resource costs that are passed on as higher prices are likely to have a regressive incidence on balance—like a sales or excise tax. By regressive we mean that lower income individuals will pay a higher *proportion* of their income in the form of these costs than will upper income individuals. This is because lower income people tend to spend a higher proportion of their income on goods and services in general. The incidence of the tax costs depends on what form of tax is used to raise the revenues. Local governments generally raise revenues by regressive sales and property taxes,[8] or equally regressive sewer charges based on water consumption or the size of the lot. The incidence of the real factor income costs, that is, those due to unemployment, is highly regressive in that the people who bear them have no income.

It is possible to design public policies to redistribute or shift some of these pollution control costs in the interest of greater equity in income distribution. There are two broad categories of cost shifting policies. The first deals with resource costs. It consists of various kinds of direct and indirect grants and subsidies designed to shift the costs from consumers as a group to taxpayers as a group, and within the group of taxpayers from lower income to higher income taxpayers. We call this a policy of *cost subsidy*.

---

[8] The extent of the regressivity of the property is not a settled issue. See Mason Gaffney, "The Property Tax is Progressive," *National Tax Association Papers and Proceedings,* 1971.

The second type of policy deals with real factor income costs, that is, those costs that arise from the slow or imperfect adjustment of resources, principally labor, to the changed economic conditions brought about by pollution control. Recognizing the parallel with similar provisions in our foreign trade laws, let us call this *adjustment assistance*. Both of these are discussed below.

Cost Subsidies

Cost subsidies can be either direct or indirect. Indirect cost subsidies can be accomplished by linking favorable tax treatment to certain kinds of pollution control activities. For example, the real cost of installing pollution control equipment can be reduced by permitting an accelerated depreciation of that equipment for corporate income tax purposes. Also, pollution control equipment can be exempted from sales taxes or real property taxes. Both types of indirect subsidy are now used by the federal and state governments. Direct subsidies consist of cash payments to municipalities and firms to reimburse them for some part of their pollution control costs. In Chapter 6 we described the federal law that authorizes grants to municipalities to cover up to 55 percent of the construction costs for water pollution treatment plants. States are encouraged by this law to put up another 25 percent— reducing the locally borne costs to 20 percent of the total.

Who does this subsidy benefit? If the state and federal tax systems are more progressive than the local taxes used to finance the town's share (which they normally are), this shifting of costs benefits lower income taxpayers at the expense of higher income groups. If the municipality receiving the subsidy is also treating industrial wastes in its plant, it is required by federal law to recover only 20 percent of the cost of treating the industrial wastes from the companies. In this way, federal funds can be used to subsidize the control of industrial pollution. The benefit of this subsidy flows to the company and perhaps to consumers in the form of a lower price. To the extent that the latter is true, the subsidy tends to favor the lower income end of the scale.

However, before one advocates more widespread use of cost subsidization policies, even on distributional grounds, there are two important qualifications to be discussed. First, although cost subsidies can change the distribution of the burden between rich and poor, they also increase the total burden because they distort the economic incentives faced by dischargers. If dischargers are relieved of the responsibility of paying the full costs of treating their wastes, they have no economic incentive to reduce or control the volume of wastes being generated. The total of wastes to be treated will be too high and the total cost of a given level of pollution control will be higher than necessary. This is true whether the cost subsidy is going to municipalities or industry. Whatever equity benefits are gained by cost subsidies come at the expense of higher than necessary treatment costs. One must always ask the question: Are the equity gains worth the cost?

As for the second qualification, where cost subsidies go to firms, there is some likelihood that their benefits will not be passed on to consumers in the form of lower prices. For example, if a multiplant firm selling in a national market receives a subsidy for pollution equipment installed in one plant, this is not likely to affect the price at which the product is sold. Hence, the benefits would accrue to stockholders. Cost subsidies can also produce interregional transfers. For example, a state subsidy to a municipal treatment plant which is also treating industrial wastes could flow largely to out-of-state consumers or stockholders if the industrial discharger was producing for a national market or if stock ownership was spread across the nation.

In short, the matter of who benefits from existing cost subsidy programs presents a decidedly mixed picture; and it is very difficult to frame subsidy policies that do not result in gross inefficiencies. The latter may easily outweigh any intended redistributive effects. Moreover, as we have noted, cost subsidies may partly go to increase profits. Accordingly, although we believe that who bears the costs of pollution control is an important policy question, we take a dim view of present efforts at cost subsidy since it is not clear that any major distributional or equity benefits are being realized. However, we would not oppose cost subsidies that were clearly targeted on needy groups—an example might be the costs of automotive air pollution control systems for low income families.

Adjustment Assistance

Adjustment assistance consists of direct payments or other contributions to specific owners of factor inputs adversely affected by environmental control policy for the purpose of shifting the burden of adjustment to newly imposed environmental standards. This type of assistance is sometimes called "targeted" assistance. Although such payments are not presently part of any state or federal environmental policy, they appear to be both efficient and equitable.

If cost subsidies are ruled out, as we argue they should be, high pollution industries will be faced with rising costs, rising relative prices for their products, and shrinking markets. Some firms would be forced out of the industry, plants would be shut down, and labor and capital would be at least temporarily unemployed. The consequences could be particularly severe in those areas where factor inputs are relatively immobile, for example, in small towns and one-industry mill towns.

An appropriately designed *adjustment assistance* policy could go far toward ameliorating the adverse effects of the pollution control policy on labor and capital that would otherwise be unemployed for extended periods. There is a useful model for this in certain provisions of the Trade Expansion Act of 1962 and the Canadian-American Automotive Trade Agreement. Both of these agreements contain provisions for adjustment assistance where the lowering of tariffs has brought about economic hardship. According to the Trade Expansion Act, it must be shown that the tariff reductions were a "major" cause of idle facilities, lack of profits, or unemployment. If this can be established, then adjustment assistance is available both for labor and for business. Firms can obtain technical assistance in developing new products or lowering costs; they can obtain low-interest loans or loan guarantees for new equipment or conversion to a new activity where market conditions are better. Workers can obtain unemployment compensation and relocation allowances for moving to areas where the prospects of employment are better. Also workers are eligible for retraining programs and grants to support them while they learn new skills.

In addition to adjustment assistance for labor and capital, a well-conceived pollution-control adjustment-assistance program should make provision for financial aid to those towns that lose tax revenues because of the loss of industry. There is a very delicate problem in defining the conditions of eligibility and in providing for an appropriate body to judge that eligibility. Yet, experience with adjustment assistance for tariff cuts should make it possible to write regulations that are neither so strict as to defeat the intent of the program nor so liberal as to turn the program into a general subsidy for business and labor.

In contrast to our conclusions on cost subsidies, adjustment assistance policies are more likely to be consistent with achieving pollution control at least cost. Furthermore, adjustment assistance deals directly with the problem of easing the plight of those who are by circumstance forced to bear a disproportionate share of the total cost of pollution control. Adjustment-assistance serves to redistribute the costs of environmental improvement where these costs are borne by a few for the benefit of many.

TECHNOLOGICAL CHANGE AND POLLUTION

We have mentioned the implications of technological change in several contexts. But its larger impact on the environment, and its potential for environmental protection deserves special discussion. Clearly, the application of advanced technology in the pursuit of high levels of material welfare has had a vast impact on the natural environment. To some extent this is inevitable. The process of agricultural development has involved great simplification of ecosystems with the incidental sacrifice of some of their natural resiliency. The result has been vastly greater production of desired plants and animals than would be imaginable under natural conditions, but the trend of natural evolution toward the building of complex, integrated, stable ecosystems has been reversed on a substantial scale.

On a similar grand scale are the possible global and regional effects of energy conversion on climates, the effect of persistent toxic chemicals on ecosystems, and the possible large-scale impact of a new technology like the SST—problems we described in Chapter 3. Generally it seems fair to say that historically, virtually no attention has been given to the environmental impacts of technological change. The development of the internal combustion engine is a recent good example, to which we shall return. There can be little doubt that Western man has taken considerable delight in his mastery over nature, propelled by science and technology. Many writers have equated the unique technological success of Western society with the religious roots of that society, roots that cleanly separate man from his environment in the development of philosophical concepts. Other writers have held this view and reinforced it with the idea that westerners are somehow uniquely greedy and shortsighted. The combination, it can be argued, is truly devastating, first to nature, then, in due course, to man.

Our view differs from this. While Western man's will for dominance has had many tragic consequences in history, it has also provided him with science, technology, and a scientific method that presumably could be redirected as a powerful force for understanding how the environment functions and for helping to reconcile, at least for a long time to come, the present conflict between high levels of living by large masses of people and a high quality natural environment.

Moreover, it is a Western idea (thought up by the Greeks, lost, and then rediscovered by the 18th and 19th century political economists) that man can design institutions to serve the goals and ends of society. These institutions are thought to be legitimized not so much by custom and religion as by the choice of society itself. Our current situation does not demonstrate that this idea was wrong, only that we have failed to practice the science and art of institutional design over a long period of time through which change was rapid and during which parts of our institutional structure were becoming rapidly obsolete.

This is the theme that has been developed throughout the book. But what does it have to do with technology and environment? Just that our system of private property provides a strong incentive for the development of a powerful technology useful in the production of private goods. The benefit of reducing costs of production or developing a good for which there is a market can be recovered, at least in large measure, by the firm, and the potential for gain provides a powerful incentive for research and development.

On the other hand, there is no built-in private incentive to develop technology that protects and enhances commonly held resources or even considers the environmental consequences of the development of technology with respect to private goods. Since the use of these valuable resources is unpriced in our exchange system, there is no incentive to conserve or improve them in either the static or dynamic context.

An example can be drawn from the automobile industry. Until recently, the industry spent virtually nothing on emission control while spending huge amounts on product development, or at least product change. Very recently, under considerable public pressure, expenditures have begun to climb. Industry outlays for research and development are not made public, but some hints are available. One is a $7 million program on emission control carried out jointly over two and one-half years by six petroleum firms and one auto manufacturer. Another is a three-year $10 million research program jointly undertaken by the Automobile Manufacturers Association and the American Petroleum Institute. In addition, companies are conducting individual research. Henry Ford announced that his company would spend over $30 million on vehicle pollution control in 1970. Suppose that industry expenditures on air pollution should rise to $50 to $75 million per year. This would still be only a very small fraction of the $1 to $2 billion a year spent on model changes.

Even if a good system of pricing commonly held resources and otherwise regulating them is worked out, thus providing an incentive to pertinent research and development (R&D) as well as static economic efficiency, this does not mean that the whole R&D job should be left to private industry. For one thing, there are broad general benefits from research. Much of the benefits of more basic forms of research are in the form of public goods. Public financial support is necessary to get the research done. In addition, Congress and other representatives of the public should not be entirely dependent on industry for information concerning technological options. Industry may have strong incentives for providing incomplete or misleading information. In an appearance before Senator Muskie's Committee, a major representative of the automobile industry gave testimony on steam engines in which he emphasized dangers of explosion and slow start-up times—problems that had been solved for some time in engines already running when the testimony was presented. Fortunately, several outside students of steam power were present and were able to correct the misimpressions this might have left.[9]

Even government contributions to R&D for dealing with environmental problems have been small relative to support for private or defense-oriented R&D. By a very broad definition of research that might be "applicable" to a range of environmental problems much broader than we have considered, only about 15 percent of the 1968 federal R&D budget would have qualified and of this only a very small proportion was actually pointed at environmental problems—one might guess less than 5 percent. This reflects the continuing preoccupation of the government with the private sector and defense.

In addition, it seems important to develop means of identifying potential environmental effects of technological development before they occur. Means should be developed to impose more of the burden for identifying such effects upon the proposed user of environmental resources. In many instances at the present time, we do not even know *what* the user is using the resource for. Fortunately, there is a strong current interest in the government and various private scientific groups in the impacts of technological change. Much of the discussion of this matter is being carried on under the rubric "technology assessment." One may hope that improved public policies will be the outcome.

POPULATION, ECONOMIC GROWTH, AND THE ENVIRONMENT

The View from the United States

There are important relationships between pollution and the size and distribution of population. However, these relationships

---

[9] *Automobile Steam Engine and Other External Combustion Engines,* Joint Hearings before the Committee on Commerce and the Subcommittee on Air and Water Pollution of the Committee on Public Works, United States Senate, 90th Cong., 2nd sess., May 27–28, 1969, Serial No. 90–82 Washington, D.C.: GPO, 1968.

are often seriously oversimplified. For example, it is not true that the increase in pollution in the United States in recent years is in anything like direct proportion to population growth. The connection between increases in pollution and increased per capita income has been much stronger. Hans Landsberg calculated that 90 percent of the increase in the electric power generation in the last thirty years has been caused by higher per capita consumption and only 10 percent by population growth.[10] The higher per capita consumption is largely due to rising income levels and the falling relative price of electricity.

A similar point can be made in connection with energy conversion as a whole, although the figures are not as dramatic. Overall energy conversion, including the portions used for space heating and industry and especially in transportation, has been increasing at a rate of more than 4 percent a year, while population increase has been nearer 1 percent. Similarly, the rise of beef consumption in the past two decades would have been about 35 percent if population growth were the only factor. But it grew 120 percent because per capita use went up 75 percent. A related illustration is the overcrowding of portions of our national parks. Population has increased by about 50 percent in the last thirty years, while attendance at national parks has increased by more than 400 percent.

In all of these cases, income growth has played a major role. The main source of growth in income in the United States has not been rising population and labor force but the increase in productivity of the labor force. For more than a century, output per worker hour has risen at a rate of between 2 and 3 percent a year. If output keeps climbing at the more recent rate of 3 percent a year, and we continue to have growth in the labor force of around 1 percent a year, measured national product will grow at about 4 percent a year for at least a couple of decades.[11] This implies that the level of measured GNP in 1980 will most likely be about 50 percent higher than it is now. Of course, unless we implement drastically improved controls, environmental pollution associated with such growth will probably be intolerable. An alternative that appeals to many is to halt economic growth.

The only way to halt growth quickly would be to reduce labor inputs either through unemployment or increased leisure. The median annual income of American households is now more than $8500. But for those earning less, our present situation is really

---

[10] In "A Disposable Feast," *Resources,* Resources for the Future, No. 34 (June 1970).

[11] Some economists have come to suspect that the growth in labor productivity is slackening.

not one of terribly great affluence. Thus it seems unlikely that there would be large increases in voluntary leisure under circumstances where a major portion of the labor force still lives at comparatively low levels of affluence. And as a means of controlling pollution, the alternative of forced unemployment is unthinkable.

Halting economic growth could produce other difficulties as well. One of these would be the lack of financial resources with which to fund environmental improvements. With continued growth, a substantial portion of the increase in our capacity to produce could be devoted to improving and protecting environmental quality rather than to producing more conventional goods. Although growth by itself is unlikely to lead equity in income distribution and better public services, it is easier to accomplish the necessary redirection and redistribution of resources when the size of the pie is growing.

We have made a case that continued economic growth is *desirable* on the economic grounds of making possible such things as greater equity. But we must ask whether continued growth is *possible* on a finite earth. In the short run (thirty to fifty years) and in the United States, there is no doubt that the answer is yes. The most serious and pressing pollution problems in this country can be controlled, if not cured without overwhelming cost, at least relative to our capacity. If there are limits to economic growth, they lie not within the borders of the United States but at the global level in the form of global pollution and limits on global resource availability. We will return to this question after a digression concerning population in the United States.

ZPG for the United States?

The population of the United States is today a little over 207 million, compared with a world population over 3½ billion, and it has increased by a little more than 50 percent since 1940.[12] It seems probable that it will increase by perhaps another 40 to 50 percent in the thirty years before the end of the century while the world population will have at least doubled. By world standards, this rate of increase is not particularly fast, but it obviously cannot continue indefinitely. The only sort of imaginable long-term equilibrium involves a rate of population growth of zero. The main question is when and under what circumstances will such a rate be achieved.

---

[12] The following discussion is based largely on an excellent article by Ansley J. Coale, "Man and his Environment," *Science, 170* (3954) (October 9, 1970), 132–136.

Many are now asserting that this ZPG equilibrium should be achieved as rapidly as possible. But if the proximate causes of pollution in the United States are high levels of production, consumption, and throughput and a lack of effective incentives for pollution control, rather than population, per se, then ZPG cannot be justified on environmental grounds. Furthermore, it appears that those who argue for immediate ZPG underestimate the problems and disruptive effects of achieving that objective. The main reason is the age composition of the present population. The postwar baby boom has left us with a very youthful society. The death rate is much lower than it would be in a population that had long had low fertility, because a high proportion of our population is in those age groups where the risk of mortality is small. Consequently, if we wish to attain a zero growth rate immediately, it would be necessary immediately to cut the birth rate in half. In other words, over the next fifteen or twenty years, women would have to bear children at a rate that would produce only a little over one child per completed family. Leaving aside the ways and means of achieving such a situation, at the end of that time we would have a very odd population distribution skewed sharply toward old age.

A more desirable and feasible goal would be to reduce fertility as soon as possible to a level where just enough children are produced to assure that each generation exactly replaces itself. However, even if this were to occur, the number of families having children would continue to increase because of the large number of children today. Therefore, population would increase 35 to 40 percent before it stabilized. The final result of the stabilization process would be that the number of persons under twenty would be about the same as today, but the number of people above that age would be substantially higher. Assuming present mortality levels persist, this stationary population would be much older on the average than any the United States has ever experienced. It would have more people over sixty than under fifteen, and the median age would be thirty-seven rather than twenty-seven as is the case today. This would be an age distribution something like that in present-day St. Petersburg, Florida—a well-known haven for retired people.

This age distribution would no longer conform to our traditional social structure—to the distribution of privileges and responsibilities in the society. In a growing population, the relatively smaller number of people at higher ages and the smaller number of high positions relative to low positions in the economy and the society tend to match up. In a stationary population, there would be a much lower expectation of advancement in professional status and career as a person moves through life. The higher rungs on the ladder would be much more crowded than at present. A population of this kind would seem to have—for better or for worse—a kind of built-in conservatism. It would seem that a careful study of the labor force, social services, and social-psychological implications of such an age structure is badly needed.

With the present marriage, fertility, and mortality patterns, couples must have an average of about 2.25 children to replace themselves with no population growth. At present, the average completed family size is about 3.0. A reduction in reproduction rates can be achieved either by increasing a family's ability to avoid unwanted children, or by bringing about a reduction in desired family size, or both. With improved education and clinical services and the development of improved contraceptive techniques, many unwanted pregnancies could be avoided. There is some disagreement among demographers as to how big a difference it would make if all unwanted pregnancies were avoided, but birth rates *would* drop substantially. One survey[13] concludes that in the period 1960 to 1965 nearly 20 percent of all births were unwanted. Elimination of all unwanted births would reduce the average completed family size from the observed 3.0 to a hypothetical 2.5.

Whether because of spreading knowledge of birth control techniques and easier access to legal abortion, or because of changing attitudes toward family size, birth rates are presently quite low and falling further. Whether this is a temporary phenomenon or represents a long-term trend is difficult to say, but some demographers are now predicting that the United States is already headed toward ZPG and will achieve a replacement level birth rate within a few years.

In order to determine whether the recent reductions in birth rates are temporary or a trend, we need a better understanding of the forces that influence birth rates. Economists have had some success in viewing the family's choice of how many children to have as an economic one. In other words, parents are viewed as weighing the costs and gains to them of having additional children.[14] One way to influence family size is to use public policy to alter the costs and gains of having children.

One possibility is to provide more equal educational and work

[13] See Larry Bumpass and Charles F. Westoff, "The 'Perfect Contraceptive' Population," *Science*, 169 (3951) (September 18, 1970), 1177–1182.

[14] See Richard A. Easterlin, "Population," in Neil W. Chamberlain, ed., *Contemporary Economic Issues*, Homewood: Irwin, 1969; "Does Human Fertility Adjust to the Environment," *American Economic Review*, 61 (2) (May 1971), 399–407; and Glen G. Cain, "Issues in the Economics of a Population Policy for the United States," *American Economic Review*, 61 (2) (May 1971), 408–418.

opportunities for women. The effect of this policy on population growth relates to the fact that one component of the cost (to the mother at least) of having a child is the foregone opportunity for creative activity and interesting work. Thus, as opportunities outside the home grow, the opportunity cost to women of having more children rises. Increased employment opportunities for women, therefore, are likely to bring about a reduction in desired family size. In a similar vein, it has also been suggested that a tax should be levied on children as a disincentive to having further children and, in fact, to reflect the social costs of additions to the population.

If economic incentives are to be considered as tools of population policy, we first need to undertake a thorough reexamination of the pattern of incentives that has developed over the long period of time during which rapid growth of both population and the economy were regarded as unquestionably socially desirable. In one way or another, society picks up a large part of the cost of having children. The income transfers from single persons and childless couples to large families because of tax laws and welfare and education policies are substantial.

If there is no genuine social value in population expansion, there is no rationale for shifting the burden, or a major part of it, to the general taxpayer (except the vexing difficulty of avoiding undesirable effects on the children that are actually born). In this light, one could think of such things as repealing income tax exemptions for dependent children, altering income-splitting provisions for tax purposes, and obtaining at least a substantial portion of the revenues to finance public education from tuition payments. Many may regard these as heresies, but they seem to deserve thoughtful consideration. Of course, how to do these things without disadvantaging the already disadvantaged child even further is an extremely serious problem. Here, as in so many areas, efforts to rationalize the system in the direction of greater overall efficiency founder on the shoals of our unwillingness as a society to come directly to grips with our income distribution problems.

The Limits to Growth: A Global View

We turn now to a brief examination of the limits to economic growth on a finite earth. We cannot hope to do justice to the subject, but only to point out the major issues and to suggest some ways in which the economist's way of looking at short-term, regional pollution problems can also be extended to encompass long-term global environmental problems.[15]

The arithmetic of exponential growth can always be made to produce startling results. Exponential growth occurs when something increases by the same percentage every period. For example, if your bank balance is growing at 10 percent per year, and it starts at $100, after one year it will be (1.10) × $100 = $110; in two years, $(1.10)^2$ × $100 = $121; in three years, $(1.10)^3$ × $100 = $133, and so on. Any exponential growth rate can also be expressed in terms of its doubling time. A growth rate of 100 percent per year means a doubling time of one year. Even relatively slow growth rates can add up to a doubling over a relatively short period of time.

Percentage Rate of Growth per Year	Doubling Time (Years)
10	7
7	10
5	14
2	35
1	70

It is a fact of arithmetic that anything growing at a constant percentage rate will grow to be very large sooner or later—sooner with a faster growth rate, or later if it starts from a very small base, but always eventually.

The limits to uncontrolled growth of any human activity, be it population, pollution, or frisbee purchases, can be demonstrated by calculating its exponential growth path over a long enough period of time. Sooner or later the exponential growth will collide with some physical constraint, such as space for people, and the point has been made.

What are the implications of the arithmetic of exponential growth for man? Unlimited growth in population is clearly impossible simply in terms of space, if nothing else. But long before population growth brings about physical crowding on a global scale, other constraints related to the capacity of the earth to provide food and resources and to absorb wastes will have been

[15] For somewhat different perspectives on the global environmental and population problems, see Paul R. Ehrlich and Anne H. Ehrlich, *Population, Resources, Environment: Issues in Human Ecology*, 2nd Edition, San Francisco: W. H. Freeman, 1972; Committee on Resources and Man of the National Academy of Sciences—National Research Council, *Resources and Man*, San Francisco: W. H. Freeman, 1969; and Donella H. Meadows et al., *The Limits to Growth*, New York: Universe Books, 1972.

met. What these constraints will be, how soon they will be reached, and what consequences can be foreseen depend in part on the pattern that growth takes. Is it a growing population with a constant per capita income and consumption pattern? Or is it a constant population with growing per capita incomes? Or is it somewhere in between?

Consider population alone for a moment. The world's population is about 3.6 billion and is expected to double by the year 2000. Even if all the families moved quickly to limit births to an average of 2.25 per family, the rate that would allow each generation just to replace itself, world population is likely to double anyway as today's children grow up to have families. A tripling of the world's population seems inevitable unless war, famine, or ecological disaster intervene. This leads to three observations.

First, this population growth is largely a phenomenon of the poorer nations of the world. The United States and other developed nations are making a relatively small contribution to the problem. They constitute a small fraction of the world's total population, and they are growing well below the world's average. During the period when the world population will be doubling, the United States population will increase by perhaps 40 percent. But if the United States achieved ZPG today, the doubling of the world population would be postponed by only one year. Thus it appears that whether or not we rapidly achieve ZPG in this country makes little difference either to us or the rest of the world.

Second, present rates of population growth impose incredible burdens on the poor countries and are a major barrier to their achieving a higher standard of living. Independent of any ultimate global environmental problem, there is a very strong case for limiting population growth as soon as possible in these countries.

Third, if and when population growth is dampened in the poorer nations and they begin to experience more rapidly rising per capita incomes, the burden man imposes on the environment may rise more rapidly than before as resource use and throughput grow at faster rates.

There have been a number of studies that have attempted to assess the capacity of the earth to support the present and expected future activities of man. The problems in conducting such studies are very difficult, and the results are controversial and conflicting. One study, sponsored by the National Academy of Sciences, had as its goal the assessment of the ultimate carrying capacity of the planet, in terms of population and its demands on available resources.[16] The conclusions covered the three broad

[16] National Academy of Sciences, op. cit.

areas of food, minerals, and energy. As for food, the Committee placed the ultimate carrying capacity at 30 billion people living at or near the starvation level. The limiting factors are the fixed input of solar energy, the efficiency of the biological system in converting that to chemical energy and protein, and the economic efficiency of man in harvesting this supply. The Committee suggested that a population of about 10 billion could be adequately fed, but went on to state

> "It is our judgment that a human population less than the present one would offer the best hope for comfortable living for our descendants, long duration for the species, and the preservation of environmental quality."[17]

The committee described its basic mode of analysis as follows.

> "Since resources are finite, then, as population increases, the ratio of resources to man must eventually fall to an unacceptable level. This is the crux of the Malthusian dilemma, often evaded, but never invalidated."[18]

But the ratio is not so simple. Two other factors must be included, patterns of demand for goods and services, and the state of technology. The quantity of resources can only be defined and evaluated relative to a given pattern of economic demand and technology. Demand and technology have combined in the past to *increase* the resource base as measured in economic terms, and this despite the tremendous demands placed on these resources by industrialization and 20th century economic growth.

The evidence to support this contention comes from a careful study by Barnett and Morse.[19] They reasoned that if resources were becoming scarce relative to the demands being placed on them, that is, if the ratio of resources to man were falling, this scarcity would be reflected in a rising price or cost for resources relative to other goods. Yet their search of the data over the past eighty years showed that in most cases the trends in prices and costs for resources were downward. We cannot conclude from this that the real cost of resources will continue to fall and that scarcity in the economic sense will never appear. But Barnett and Morse's explanations as to why scarcity has not already occurred should help us to understand what is likely to happen when the trend reverses and scarcity increases.

[17] Ibid., p. 5.
[18] Ibid., p. 8.
[19] Harold J. Barnett and Chandler Morse, *Scarcity and Growth: The Economics of National Resource Availability,* Baltimore: The Johns Hopkins Press, 1963.

If a resource becomes scarce relative to the demand for it, its price will rise. This triggers a number of economic responses, all of which help to mitigate the scarcity. Less rich or productive sources can now be economically developed; there is an incentive to search for new sources of the resource in the case of minerals; techniques for economizing on the use of the resource and getting more per unit are adopted; there can be substitution of other resources; the prices of the products containing the resource will rise, shifting consumption toward other more abundant goods and services; there may be increased recycling and materials recovery; and, there will be research to develop new technologies to do all of these things better. There are two important characteristics of this set of responses. First, the bigger the price increase, the bigger the incentive and the larger the responses. Second, the responses tend to be graduated and continuous.

The most notable feature of the economic history of Western man is the high rate of technological discovery and resource use per capita that have resulted in the current high standard of living. The present period could be characterized as one of relative resource abundance per capita or, conversely, labor scarcity per unit of resources. Looking to the future, whether innovation will be rapid enough to keep resources relatively abundant in the economic sense is an open question. Intuitively one would expect that the answer is "No." But in that case, the other economic mechanism will be brought into action. If technological change is not rapid enough to keep ahead of physical resource scarcity, the cost of high-resource-using goods will rise. The past trend of economizing on labor by high rates of resource use will be reversed. There will be a trend toward the production of services and highly durable goods and away from resource intensive forms of consumption. Recycling of materials will expand. Over the long run, this would require very great changes in the composition of output and even in the way of life. But would this be a greater change than that which has occurred over the last, say, 500 years?

On the basis of this discussion, what kind of advice can we give as to how to navigate into the future? First, as to world population, its growth cannot go on indefinitely. Projections of the population growth before ZPG is achieved are alarmingly high, even if we cannot foresee the precise consequences of these totals. More importantly, the job of *controlling* and ultimately ending population growth and the difficulty of coping with the final stabilized total get more difficult every day. Birth control efforts must be given high priority in those areas where population growth is most rapid.

Second, our global geophysical and ecological systems hold a number of dangerous traps along the way. Some of these were described in Chapter 3. Examples are global DDT poisoning and the possible irreversible effects of large numbers of SST flights. Not all of these traps are clearly perceived, nor have they all been identified and marked. Avoiding them will require considerable research to fill gaps in our knowledge of how the earth works and a well-developed sense of caution about pushing ahead with new technologies before their ramifications are fully understood.

Finally we need to do a better job of creating institutions for managing and controlling those actions of man that affect the environment. We described the kinds of economic adaptations that are triggered when rising price signals the increasing scarcity of a resource. But these signals have been absent where environmental resources are concerned. And though these resources are becoming increasingly scarce, the kinds of economic adaptations that are necessary for man's long-run welfare, if not survival, have not been taking place. We face a major task in creating the appropriate institutions and endowing them with sufficient powers of the right kind.

9

Environmental Management: An Overview

This final chapter starts with a discussion of some things this book is *not* about. Our focus has been on economic concepts because we are economists and believe that economics provides indispensable insights concerning the root causes of the environmental problem and its effective and efficient cure. But the reader who has stuck with us this far will have realized, if he did not know it already, that understanding and managing the environment is *inherently* an interdisciplinary problem. As well as using such economic concepts as the market system, common property resources, and benefit-cost analysis, we had occasion to invoke the principle of mass balance from physics and to refer to numerous concepts and analytical procedures drawn from such diverse fields as biology and meteorology. To understand fully the nature and significance of the environmental problem and its potential solutions, some knowledge of all these fields is necessary.

However, in our discussion of the economics of environmental management, we have given only incidental attention to two other features of our institutional structure that have a great, and in some cases overwhelming, bearing on our ability to confront environmental problems successfully. These are our legal and political systems. We now turn to these. This is not to say we will discuss them fully—books could be written about environmental politics and environmental law, and have been. We want to draw attention to some of the more salient aspects.

In the final section of this chapter, we undertake to draw the strands together in a schematic fashion and provide at least a glimpse of the management system in its entirety.

THE LEGAL SYSTEM AND ENVIRONMENTAL POLICY

In addition to enforcing existing laws governing pollution, legal doctrine has evolved ways of attempting to deal with pollution through the courts. For example, according to the so-called *nuisance doctrine*, an individual can sue a polluter for compensatory damages and to compel him to stop. However, to gain standing in court, the individual has to show that he is uniquely affected by the action of the polluter. If all the soot from a nearby smokestack lands on his house (or at least a small number of adjacent houses), he can sue. But if the soot is widely dispersed throughout the urban area and adversely affects many people, no one person may sue under the nuisance doctrine.

The typical pollution situation is one in which a relatively large number of parties (individuals, firms, and units of government) is damaged by one or more sources of residuals discharge. The damage cost to each individual may be small individually but large when summed over all the damaged parties. In these circumstances, a legal suit by an individual to try to remedy the situation has some of the characteristics of a public good. If he wins, not only does he gain, but all of the other damaged parties also gain—they are free riders. Accordingly, the individual party does not have an incentive to undertake the expense and unpleasantness of a legal action. Conversely, the group of damaged individuals face significant difficulties and costs in organizing for collective legal action. As a consequence, private legal actions against residuals dischargers have been rare.

The difficulty private individuals have in bringing legal action has been reinforced by the so-called *public nuisance* doctrine.[1] For those widespread public nuisances for which private nuisance actions are not possible or practical, public nuisance suits can be initiated only by public officials (for example, state attorney generals or United States attorneys) acting on behalf of all individuals in their jurisdictions. Under traditional interpretations of the doctrine, an individual does not have "standing" in these cases. This rule has been followed by the courts ever since it was set down by an English judge in 1536.

There is now a perceptible new trend toward greatly broadening the rights of individuals to seek legal redress from environmental pollution. This trend is partly the result of new laws and partly the consequence of developing judicial precedent. An example of the latter is the development of the concept of *class action*, by which an individual or organization can sue on behalf of large groups of similarly affected citizens. The Environmental Defense Fund has been a pioneer in this area by bringing a series of actions against the use of DDT.

---

[1] For a comprehensive discussion of the matters addressed in the remainder of this section, see Joseph L. Sax, *Defending the Environment*, New York: Alfred A. Knopf, 1970.

Michigan and Connecticut have recently enacted laws that empower any person or organization to sue any private or public body and to obtain a court order restraining conduct that "is likely to pollute, impair, or destroy the air, water, or other natural resources or the public trust therein."[2] According to these laws, once the plaintiff has introduced evidence of such violation, the defendant—if he is to prevail—must affirmatively prove that there is no feasible and prudent alternative to his conduct and that it is consistent with promotion of the public health, safety, and welfare. The fact that legislation so at variance with long-established practice passed is both surprising and encouraging. Apparently, large and enthusiastic public attendance at hearings, and newspaper and organized labor support was persuasive to the legislators. Similar bills have been introduced in at least five other states. And, in 1971, a bill establishing the right of citizens to sue polluters directly was introduced into the Congress.[3]

In a related development, the National Air Quality Standards Act of 1970 gives any citizen the right to sue any individual for violating an air quality standard. Even more importantly, it permits suits against public officials for not carrying out their responsibilities under the Act. However, a late amendment restricted this latter provision to those cases where the official was obligated by law to act and failed to do so. The amendment exempted from citizen review through the courts those cases where an official had discretionary authority to act.[4]

The National Environmental Policy Act (NEPA) has proven an effective legal weapon for individuals and organizations challenging environmentally harmful actions of the federal government. This Act establishes environmental objectives as national policy, requires that environmental values be integrated into and weighed with other values in all federal decisions, for example, on whether to build a highway, and requires federal agencies to compile and make public reports on the potential environmental impacts of their decisions. Many governmental actions have been challenged in court on the grounds that one or more of these provisions has not been met. While most of this litigation is still pending, it appears that NEPA will have a far-reaching impact on government decisions affecting the environment.

A still rather tentative but possibly significant new concept is

[2] The quotation is from the Michigan Law.

[3] The bill, S. 1032, was introduced by Senators Hart and McGovern.

[4] The Federal Water Pollution Control Act passed by the Senate in 1972 (see Chapter 6) bestows on citizens the standing to bring both industry and the EPA to court when they believe standards are not being met.

presently being developed in the courts. It is based on an ancient legal theory known as "*the public trust.*" Its origin is in Roman law, and it is based on the idea that certain common properties are held by government in trusteeship for the use of the general public. This idea penetrated into Anglo-American law but has been seldom applied except to a few sorts of public properties.

There is now renewed interest in the concept that rests on three related principles. The first principle is that certain resources —such as the air and the sea—have such importance to the citizenry as a whole that it would be unwise to make them the subject of private ownership. The second is that these resources partake so much of the bounty of nature, rather than individual enterprise, that they should be made freely available to the entire citizenry without regard to economic status. The final principle is that it is a basic purpose of government to promote the interests of the general public rather than to redistribute public goods from broad public uses to restricted private benefit.

The most famous public trust doctrine case in United States history occurred in 1869 when the U. S. Supreme Court reversed the deeding of a mile of Chicago lakefront to the Illinois Central Railroad by the State of Illinois. The Court agreed that the state held the title, but ". . . it is a title held in trust for the people of the state. . . ." The public trust doctrine has been brought to bear in a limited way on contemporary problems by requiring users of public properties to explore a greater range of options than they normally would have done. Some legal experts believe broader applications to common property resources are in the cards.

Perhaps the most sweeping extension of legal rights and powers on behalf of the people is embodied in the many proposals for an environmental bill of rights. Many state legislatures are or will be considering constitutional amendments that would make a clean, healthful, and reasonably quiet environment a matter of constitutional right. It must be realized, however, that the establishment of a constitutional guarantee of a clean environment will not change things overnight. Such a guarantee would be like the Supreme Court decision of 1954 banning segregation in schools. It would promise a lot but would not in itself contain the means for delivering on the promises. The holders of the guarantee would still have to rely on the state as the instrument of collective action to enforce the guarantee.

These legal developments suggest a gradual shifting of the burden of proof away from damaged parties and toward those who would damage the environment. This is likely to make recourse to legal action more frequent than in the past—and more successful. The shifting in burden of proof is also likely to lead to a more careful examination of possible effects on common property resources when a new technology is developed and applied.

While these developments are welcomed by those interested in environmental improvement, there are some important limitations to the legal approach to massive environmental quality problems. First, by its nature, the legal approach is piecemeal, dealing with individual polluters and individual problems. Also, the adversary climate of the courtroom may not be conducive to working out comprehensive and efficient plans for managing environmental resources. In the cases of private and public nuisance suits and class actions, the burden of proving damages or adverse effects generally lies with the individual, and proof can be difficult. For example, the scientific evidence regarding the adverse health effects of air pollution may not be up to legal standards of proof. More importantly, preparing such suits, gathering information and evidence, and preparing the necessary expert testimony is very costly, and the funds must be raised from private sources.

The question of the self-sustaining nature of such efforts is an important issue. If the legal strategy is to be more than an effort to seek out and publicize a few selected examples, it must have the resources to adopt a more comprehensive approach to the problem. As the recent Nader Report on federal water pollution control policy stated: "If the citizen effort is to be a viable *countervailing* force against polluters, it . . . must have money of its own . . . to support lawsuits and purchase technical assistance comparable to that which corporations can buy."[5] Today no mechanism exists to insure that such citizen efforts will be more than sporadic efforts dependent on charity. Proposals that citizen legal efforts share the fines imposed on polluters with the government could go some distance toward remedying this problem of coverage and self-sustenance, especially if the fines imposed are sufficiently high.

With these constraints on the operation of a legal strategy, perhaps the most useful contribution of such an approach would be to serve as a vehicle for keeping both public officials and private individuals on their toes by forcing them to consider all of the consequences of their particular actions, and subjecting their actions (and failures to act) to careful review and possible reversal. As such, legal actions are best viewed as supplements to more general management tools, such as effluent charges.

---

[5] David Zwick and Marcy Benstock, eds., *Water Wasteland, The Ralph Nader Study Group Report on Water Pollution*, New York: Grossman, 1971, p. 397.

THE POLITICS OF POLLUTION

In discussing the problem of environmental pollution, we have treated the environment as a common property resource for which the normal incentive structure of the market system has failed. We have examined several alternative institutional arrangements for remedying this situation. One of these—the creation of economic incentives through the imposition of residuals charges—has been viewed as being generally superior. Although this is a widely (but not unanimously) held view among economists, it has not yet attracted very much political support. An examination of the political process and its operation with respect to environmental problems will help us understand why this is so.

The political system exists in large part to reconcile or resolve conflicting interests. Many of these "interests" are economic in nature. With respect to the pollution issue, the conflict is over how much cleaning up of pollution is actually going to be done, and who is going to pay for it. In a more fundamental sense, the conflict is over the unresolved property rights in the environment. Do these rights reside in those whose use of the environment for discharging residuals reduces its quality and the flow of other services that it yields? Or do they reside in those whose use of the environment for other purposes is benign and who suffer losses from its deteriorated quality?

In trying to gain a better understanding of how our political system has responded to the pollution issue, we will explore some simple and informal propositions concerning such a pluralistic political system. These propositions concern both the policy makers concerned with carrying out or enforcing pollution control policies and the places in the political structure where such decisions are made.

The first proposition is that policy makers, like other decision makers, act in their own self-interest. Elected officials behave so as to assure that they will be reelected; bureaucrats behave so as not to alienate the legislature or to assure themselves a lucrative position in private industry at a later date. If a policy maker supports a policy that takes something away from a part of his constituency, he alienates that group. As a consequence, he would tend to support the policy only if the losers were small in number or without influence, and if the gainers from the policy were large in number or influential. Thus, the second proposition is that policy makers search for policies whose costs are hidden or can be shifted to less influential elements of their constituencies. Third, policy makers try to postpone decisions (since every decision has a cost) and to avoid the costs of a decision by shifting the responsibility for making it to another place in the political system.

In evaluating the political structure implicit in these propositions, two criteria are relevant: accountability and accessibility. In a system with full information on the part of both policy makers and voters and with no basic structural imperfections, a decision maker would be accountable to all groups affected by his decisions, and all of these groups would have access to him.

However, such perfection does not characterize our political system. Institutional arrangements, such as the Congressional seniority system and loose campaign financial regulations that enable economic power to be readily transformed into political power, undermine both accountability and accessibility. When such structural imperfections persist, policy makers gain the leeway that they desire to avoid those decisions likely to alienate major sources of their support. They have the latitude to choose the strategies that shift costs to unwary sectors of the constituencies, that shift responsibility for decisions that are unpopular with some group to other parts of the political system, and that postpone the decisions that are likely to generate adverse political repercussions.

Given the incentives implicit in these propositions, we can gain some insight into why the type of pollution control policy that has emerged from our political process is not unexpected. Let us review some of the characteristics of existing policy.

1. Through existing and pending legislation, federal legislators have consistently shifted the burden for making difficult decisions to the states (for example, the setting of standards), or to decision makers within the federal bureaucracy. As a result, those who make these decisions are typically both less accountable and less accessible than the legislators who pass the basic legislation. Federal legislation typically passes with large majorities, sometimes unanimously. One can only conclude— if everybody is for it, they must have found some way to duck the real issue.

2. Policy makers have been willing to subsidize industrial dischargers wherever this could be hidden in tax depreciation formulae or municipal cost-sharing programs. By accepting a share of the financial burden of pollution control, policy makers have, in effect, assumed that dischargers have a property right in the environment.

3. Federal law requires that the states must hold public hearings before setting air quality standards. Setting standards involves finding an answer to the question: "How far do we want to

go?" Obviously there can be considerable public conflict around this issue. Public involvement through hearings is desirable so that issues can be clarified and affected parties can gain access to policy makers. Moreover, depending on who actually makes the decision, there can be greater political accountability. However, the federal air quality law enacted in 1970 requires that the Administrator of the Environmental Protection Agency set national air quality standards (and emissions standards). It eliminates the requirement for public hearings. Again, through indirect means, accessibility is reduced and those who are ultimately responsible for public policy—the legislators—have increased the difficulty for those affected to hold them accountable.

4. Federal law does not stipulate that public hearings be held in the establishment of water quality standards. Some states have established these standards legislatively while others have delegated the responsibility to regulatory agencies. In the latter case, political accountability and public access to the decision are again diminished.

5. Setting environmental quality standards is a meaningless exercise unless effective mechanisms are developed for achieving the standards. Almost without exception, states have placed primary reliance on some form of licensing of discharges accompanied by judicial enforcement of the license terms, as discussed in Chapter 5. The political aspects of this procedure are quite unfavorable to effective pollution control.[6] The adoption of a licensing system does not resolve the political conflicts inherent in pollution policy. Instead, there is a continuing political conflict over license applications, terms and limits on discharges, and enforcement. Furthermore, these battles are fought on terms that are advantageous to dischargers for several reasons. The political issue is removed from the legislative arena and placed in a bureaucratic one where accountability and accessibility are less. The choices are more likely to be framed in technical terms that make it difficult to articulate the public interest. Finally, each discharger has a lot at stake in each decision and has large incentive to devote resources and energy toward swinging the decision his way; the public interest is diffuse, and because few persons have the incentive or the command of resources, likely to be poorly represented.

As these characteristics demonstrate, the political system has tended to respond to the emerging environmental problem by

[6] See pp. 102–106.

shifting the real decisions from the federal to the state level and from legislatures to bureaucratic agencies. It has tended to make decisions in arenas where there is less accountability and accessibility, and to avoid final resolutions of the political conflicts in favor of piecemeal fragmented decisions. All of these tendencies work against the public interest in pollution control and in favor of polluters.

From this perspective, it is little wonder that the economist's prescription of residuals charges has generated modest political appeal. It goes against all of these tendencies. Establishment of a charge system in conjunction with environmental quality standards would resolve most of the political conflict over the environment in a highly visible way where those who would be hurt by such a policy could see what was happening. It is just such open and explicit choices that policy makers seek to avoid.

It may not be facetious to suggest that the reason residuals charges have not been effectively tried in this country is that they would work. In the absence of effective pollution control policies, dischargers are able to expropriate the environment for their own use by the act of discharging residuals. Such things as discharge licenses and standards at best place limits on dischargers' rights to use what they *de facto* if not *de jure* own. But a system of residuals charges establishes the principle that the environment is owned by the people who use its other services, and dischargers must buy the right to use part of the environment for waste disposal. Such massive transfers of property rights and the wealth they represent seldom occur in the absence of substantial political upheaval. The most formidable barrier to controlling pollution is probably not technology, population, or public attitudes, but the politics of power.

A RETROSPECTIVE VIEW

In this book, we have developed the basic economics of environmental quality and followed the logic of these economic principles to their policy conclusions. Fundamental to the perspective that we have developed is the idea of managing the environment so as to maximize its contribution to economic welfare. In this concluding section of this concluding chapter, let us look back over the framework we have developed. In undertaking this retrospective look, we will refer to Figure 9-1. This chart also provides a convenient means for reviewing how other disciplines besides economics fit into the overall environmental quality management

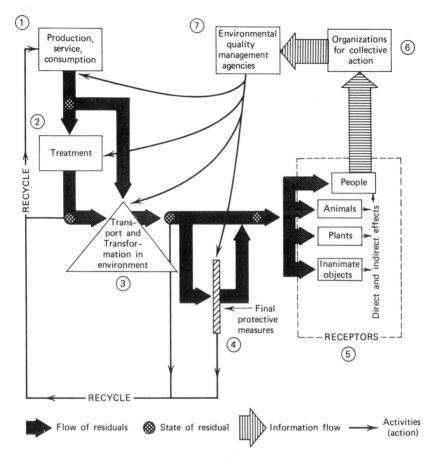

Figure 9-1 Schematic depiction of a residuals management system. *Source.* Adapted with permission from Allen V. Kneese, Robert U. Ayres, and Ralph C. d'Arge, *Economics and the Environment: A Materials Balance Approach,* Washington: Resources for the Future, Inc., 1970.

picture. The boxes in the schematic are labeled 1 to 7, and we will discuss each one briefly in turn.

The first box contains the production and consumption activities of the economy. Included here is a vast array of service and materials flows that incorporate the basic extraction of raw materials, their processing into intermediate and final products and distribution to consumers, and the return of waste flows to the environment. Most of these flows have counterpart market transactions that assign values to them and ration and conserve their use. But the residuals flows that interact with common property resources, such as watercourses, the air mantle, and large ecological systems, do not.

Some amount of residuals flow from a large modern complex economy is inevitable. To reduce the flow to zero would be incredibly expensive. Indeed, it would require an economic system in which all materials are held in closed recycling systems, and only solar energy is used. With our present knowledge, this is an impossible dream. However, this is not to say that the amounts and types of residuals flows are fixed. Indeed, they are highly variable, depending on the production and consumption technologies that are adopted. In the absence of any regulation, management, or ownership of the environment, the residuals flow is limited in size only by the rate of production of goods and services. This result is far from optimal because it takes no account of the external costs imposed through deterioration in the quality of common property resources caused by the residuals. If restrictions are placed on the use of common property resources for residuals disposal, the incentive structure will change. There will be an added incentive to reduce the flow of residual materials by means of recycling processes, the use of materials with lower waste by-products, and the substitution of production produce less harmful forms of residuals.

However, there is the possibility of altering residuals streams further after generation. This is through the complex of procedures we label "treatment" in box 2. It is important to recall from the materials balance model that treatment does not reduce the mass of residuals but only alters their form or location. In fact, the total mass is increased since the treatment processes themselves require inputs. If the disposal of only certain residuals is restricted, dischargers will respond by treatment in which some kinds of wastes are substituted for others. Hence, a policy which aspires to optimality of the whole system must place a price on (or otherwise restrict) all of the residuals streams simultaneously to get the right mix of residuals.

The disciplines that are most centrally involved in understanding the functioning of the aspects of the environmental quality management system shown in boxes 1 and 2 are economics and engineering.

Residuals that are emitted, either directly from the production and consumption processes or indirectly from treatment processes, enter a natural environmental system. It may be the land, atmosphere, watercourses, or ecological systems associated with any of these. The transport and transformation processes that occur in these natural environments are indicated in box 3. These were represented by the transformation functions of Chapter 5. Sanitary engineers, hydrologists, meteorologists, and ecologists have built intellectual idealizations or models of these transport and transformation phenomena. These models usually involve the solution of complex systems of simultaneous equations and are used extensively in research and planning activities in regard to environmental quality management. Although these models are very useful, considerable effort is required to expand the models to cover more variables and to test the ability of the models to reproduce the real world phenomenon by comparing their predictions with actual observations.

Box 4 is labeled "final protective measures." It includes such things as treatment of water prior to use for industrial and municipal purposes, the conditioning and filtering of air in buildings, and the application of corrosion-resistant paint and coatings to exposed surfaces. The market system based on individual consumer and business decisions operates rather effectively to implement these measures—clearly more effectively than it does to control residuals discharges. To be sure, an optimal (or efficient) system of environmental management would no doubt include many of these measures. However, without the effective control of residuals discharges, these final protective measures will have to be used to a larger extent than is optimal or efficient. Visualize, for example, people scurrying around in gas masks from one sealed and filtered building to another—a protective measure—while inexpensive means of controlling particulate discharge go unused.

Box 5 identifies the "receptors"—that is, those users of the common property resources who are injured by residuals discharge to the natural environmental systems. The extent of this injury is a complicated function of many factors—the amounts and types of residuals discharged, the specific transformations in the environment, the extent of use of final protective measures, the type and location of receptors, and, finally, the values people attach to the physical, chemical, and biological effects caused by residuals in the environment. In this part of the management system, economics is valuable in providing an evaluation criterion (willingness to pay) and certain methods of estimation, but many other disciplines are involved as well. These include prominently agricultural, materials, and medical sciences (including, particularly, epidemiology and genetics). To the extent that people's evaluation of damage is conditioned by the particular way they perceive what has happened to the environment, the discipline of social psychology also becomes pertinent.

In Chapter 4 we explained that environmental resources are "public goods." This means that the management of residuals discharge to them must be executed through some form of collective action. This is where box 6 and the "organizations for collective action" come in. These take a variety of forms. They may be private, such as "conservation" associations that bring legal actions, or they may be formal political bodies such as the United States Congress or state and local governments or the governing board of a river basin or airshed authority. The formal political organizations perform a legislative function. They create "public policy" of the kinds described in Chapters 6 and 7. Often the policies devised entail delegation of authority to what are essentially executive agencies. These are indicated in box 7. Examples include the Federal Environmental Protection Agency, and the various state air and water pollution control boards. In principle these agencies are to carry out public policy, not make it. In practice, however, these agencies have substantial policy discretion. As a result, policy is in reality created by a combination of legislative institutions and executive agencies. The elements of such policy are various types of standards, enforcement procedures, pricing and taxing policies, and public investment in environmental quality management facilities.

The solid lines from box 7 to boxes 1, 2, 3, and 4 show the points at which public policy can alter environmental quality. For example, charges can be placed on emissions to common property resources, inducing action in boxes 1 and 2; for instance, increasing the recycling and treatment of particular residuals streams. In some instances it is possible to influence the transport and transformation processes in the environment—hence, the line to box 3. Among the examples we have used previously is the low flow augmentation of streams and the insertion of reaeration devices. Finally, as the line to box 4 indicates, management agencies may assume a role in providing final protective measures as when a municipal government constructs and operates a water treatment plant.

Many disciplines in addition to economics are required to deal effectively with the issues implicit in boxes 6 and 7. In particular, political science, public administration, law, and sociology can contribute to an understanding of the efficient functioning of organizations for collective action and environmental quality management agencies.

Looking at the environmental quality management system as a whole reveals its appalling complexity. Many of its elements are understood only dimly. It would be desirable if this system were fully comprehended before undertaking to manage environmental quality. However, action cannot await perfect knowledge. The problems are too urgent and the political process responds—whether intelligently or blindly, representatively or with bias, it responds. Therefore, we must use the considerable insights we have gained from a more partial view of the process, though imperfect, to improve public management of environmental quality. Up to this point, as we have emphasized, public policy has not made good use of what is already known—especially what is known of the economic nature of the problem.

From our analysis, then, several steps seem to be urgently needed. Although they are not in themselves a fully optimal constellation of activities, they will move us in the right direction. These steps are:

1. Wherever possible, taxes or charges should be imposed on the use of common property resources for residuals discharge.
2. Institutions should be developed to manage river basins and airsheds on an integrated basis. These institutions should have genuine authority and at the same time be accessible and accountable to the people.
3. The functioning of the legal process should be improved, especially with respect to "public nuisance" situations.
4. Adequate government support for research on all aspects of the environmental quality management system should be provided.
5. The political process should be improved so as to increase its visibility, accessibility, accountability, and representativeness.

Although these may seem to be tough prescriptions, they are, in fact, a minimal set required for the better management of the environment. The challenge of improving environmental quality implies the need for basic and fundamental changes in our present way of doing things.

Index

Abortion, U. S. population growth and, 155

Adams, F. Gerard, 110

Adjustment assistance, pollution control and, 146, 147–148

Air pollution, 21, 45–48
automobiles and, 28, 46, 47, 48, 50, 53, 129, 132–136
control of, 52–53, 124–138, 141, 143–144, 168–169
cost-and-benefit study of, 125–132, 139, 141
federal policy on, 130–132
forests and, 20
fuels and, 137
geographic concentrations of, 50–52
human effects of, 45–48, 124–125, 126–128
lead in gasoline and, 48, 135
life expectancy and, 126–127
national emission standards and, 4–5, 131, 132–133, 134
sources of, 48–49
tax on, 133–134
types of, 45–49, 132
variation over time, 48, 50–51
water pollution and, 124–125

Air Quality Act of 1967, 130

Albedo, increase of, 41–42, 43, 44

Algae blooms, water pollution and, 56–57

American Chemical Society, 21, 45

American Petroleum Institute, 150

Atomic Energy Commission, radioactive pollutants and, 60

Automobile Manufacturers Association, 150

Automobiles, air pollution by, 28, 46, 47, 48, 50, 53

maintenance for pollution control, 132–133
pollution control and, 4–5, 28, 129, 132–136
as solid waste, 63
thermal reactor and, 135

Automotive emission pollution, government standards and, 4–5, 131, 132–133, 134

Ayres, Robert U., 14, 17, 171

Barnett, Harold J., 159

Benefit-and-cost analysis, see Cost-and-benefit analysis

Benstock, Marcy, 105, 120, 166

Biochemical oxygen demand (BOD), water pollution and, 53–54, 57

Birth control, population growth and, 155

Boulding, Kenneth, 19

Bower, Blair T., 46, 126, 142, 143

Bryson, R. A., 41

Bumpass, Larry, 155

Cain, Glen G., 155

California, carbon monoxide pollution study by, 5

Canada, water pollution in, 58

Canadian-American Automotive Trade Agreement, 148

Carbon dioxide, cumulative effects of, 41, 93
increase in quantity of, 40–41, 44

Carbon monoxide, air pollution by, 46, 49, 132
automotive standards and, 4–5

Chamberlain, Neil W., 155

Chicago, sulfur oxide pollution in, 51

Class actions, in pollution suits, 163–164, 166